Bilkent University Lecture Series

İzzet Şahin

Regenerative Inventory Systems

Operating Characteristics and Optimization

Springer-Verlag
New York Berlin Heidelberg
London Paris Tokyo

Prof. İZZET ŞAHİN
School of Business Administration
University of Wisconsin-Milwaukee
Milwaukee, Wisconsin 53201
U.S.A

Library of Congress Cataloging-in-Publication Data
Şahin, İzzet,
Regenerative inventory systems: operating characteristics / by İzzet Şahin.
p.   cm.
Includes bibliographical references.
ISBN 0-387-97134-3 (U.S.)
1. Inventory control. 2. Renewal theory. I. Title.
TS160.S24   1989
658.7'87--dc20   89-21838

Printed and bound by Meteksan A.Ş., Ankara, Turkey.

9 8 7 6 5 4 3 2 1

ISBN 0-387-97134-3 Springer-Verlag New York Berlin Heidelberg
ISBN 3-540-97134-3 Springer-Verlag Berlin Heidelberg New York

ISBN 975-7679-00-3 Bilkent University Ankara

To Murat and Tolga

Preface

This is a renewal-theoretic analysis of a class of single-item (s, S) inventory systems. Included, in a unified exposition, are both continuous and periodic review systems under fairly general random demand processes. The monograph is complete in the sense that it starts from the derivation of the time dependent and stationary distributions of basic stochastic processes related to these systems and concludes with the construction and testing of simple, distribution-free approximations for optimal control policies. However, it is rather incomplete as an account of single-item inventory systems in that it narrowly focuses on systems with full backlogging of unfilled demand and constant lead times, through what has come to be known as stationary analysis.

The level is intermediate, and the style is informal. Some prior knowledge of probability theory and inventory control is assumed on the part of the reader. Given these, the monograph is self-contained. Extensive use is made of renewal-theoretic concepts and results; these are reviewed in Chapter 2.

The text relies heavily on my previously published work on the subject. Over the years, this research has been supported by the Scientific and Technical Research Council of Turkey, National Research Council of Canada (1978-79, A3074), Management Research Center of the University of Wisconsin- Milwaukee, and the National Science Foundation (ECS-8011916). I am grateful to Dr. Diptendu Sinha of the University of Notre Dame for his very substantial help in computational work and to Dr. D. J. McConalogue of the Delft University of Technology for providing us with his algorithm and software for its implementation.

A prepublication draft of the monograph was used as lecture notes for a graduate course on inventory theory at the Industrial Engineering Department of Bilkent University. I thank my students, Zeki Akbaş, Cemal Akyel, Ayşen Eren, Levent Kandiller, Nureddin Kırkavak, Ceyda Oğuz and Hakan Polatoğlu, for discovering a large number of errors in the text.

Many thanks, also, to James Sagovic at the University of Wisconsin-Milwaukee and Konuralp Ünyelioğlu at Bilkent University for an excellent job of typing the manuscript.

Spring, 1989
Milwaukee, Wisconsin and
Ankara, Turkey

İzzet Şahin

Contents

Chapter 1

Introduction

This monograph is about a class of single-item (s, S) inventory systems that can be investigated through renewal theory. These systems are characterized by: 1) independent, identically distributed (i.i.d.) batch sizes (quantities demanded) separated by i.i.d. interdemand times, 2) full backlogging of unfilled demand, and 3) a constant procurement lead time. The first two representations give rise to a simple regenerative structure, and the constant rather than random lead times result in additional simplifications. Included are both continuous review and periodic review systems, with the latter being treated as a special case of the former.

For the continuous review system, let X_n be the length of the n-th interdemand time and Y_n the size of the n-th demand. Let $L \geq 0$ denote the length of the lead time. At time t, inventory on hand is $I(t)$ and *inventory position* is $I_p(t)$. The latter is defined as the stock on hand plus on order minus backorders. The control policy is an (s, S) policy that operates on the inventory position. Thus, whenever $I_p(t) < s$, an order of size $S - I_p(t)$ is placed which arrives L time units later. Clearly, there could be several orders outstanding during a lead time (Figure 1.1). In terms of inventory costs, we impose a simple structure on the system involving a fixed setup (ordering) cost

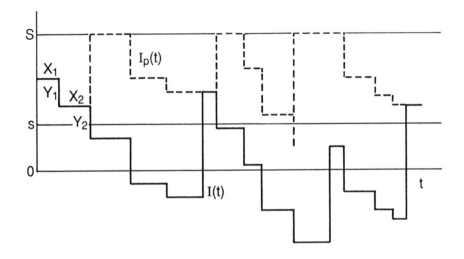

Fig.1.1. The continuous review system

K and linear holding and shortage costs with respective rates h and p (\$/unit/unit time).

The continuous review system described above was first investigated under constant lead times by Beckman (1961). Among more recent work, the most relevant ones for our purposes are Tijms (1972) and Stidham (1977, 1986). Our treatment is based in part on Sahin (1979, 1982, 1988a, 1988b) and Sahin and Sinha (1987).

For the periodic review system, let Y_n be the demand during the n-th period. The sequence $\{Y_n, \ n = 1,2,...,\}$ is i.i.d. It is assumed that the lead time is an integral multiple of the common period length which in turn is regarded as unity. The control policy is again an (s, S) policy operating on the inventory position at the end of each period. Period ends are review points. If the inventory position at a review point is $< s$, an order is placed to raise it to S. Otherwise, no action is taken. The order placed arrives after L periods. There is a fixed ordering cost K; holding and shortage costs are linear and are charged against period ending inventories at rates h and p (\$/unit/period) respectively (Figure 1.2a).

Following Arrow et al. (1951, 1958), the periodic review model has been investigated by many authors. Optimality of (s, S) type policies for the corresponding multi-period dynamic model was established by Scarf (1960). He showed that if the procurement cost is of the form $K + cz$, $z > 0$, where K is the fixed charge and c the unit cost, and if the one period expected holding plus shortage cost function is convex, then the optimal policy for period n is an (s_n, S_n) policy. Scarf's proof was extended to non-differentiable cost functions by Zabel (1962), and, an alternative proof under slightly different conditions was provided by Veinott (1966). For the stationary model of interest to us, Iglehart (1963) showed that under Scarf's assumptions, the sequences $\{s_n, n = 1, 2, ...\}$ and $\{S_n, n = 1, 2, ...\}$ converge to limit points s and S which form the optimal policy for the infinite-horizon problem. Some of this literature is reviewed below.

The periodic review model can be handled as a special case of the continuous review model introduced above (cf. Stidham, 1986). Since the inventory position is reckoned and holding and shortage costs are charged at the end of each period, we may assume without any loss that the total demand of a period is realized in one batch at the *beginning* of the period, immediately following any order receipt. The inventory position then remains the same until the end of the period. In this version of the periodic review system, an order will be placed if and only if the inventory position has been below s throughout the previous period. To reconcile this with the order-triggering mechanism of the continuous review system, we may assume, again without any loss, that an order is placed whenever the inventory position falls

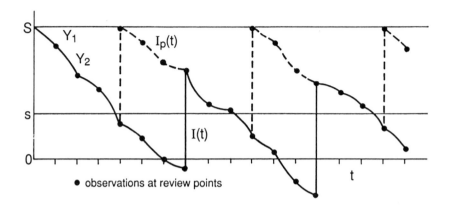

Fig.1.2a. The periodic review system

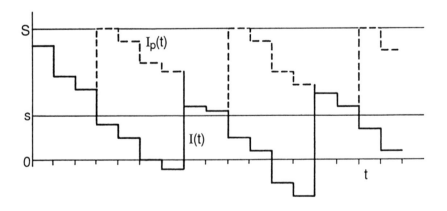

Fig.1.2b. The corresponding continuous review system with
constant interdemand times

below s but that one period is added to the original lead time. With reference to the original system, this amounts to placing the orders at the beginning, rather than at the end, of the period in which the inventory position falls below s, but adding a period to the original lead time. With these conventions, the periodic-review system is effectively equivalent to the continuous-review system with constant interdemand times (Figure 1.2b). Since in the modified system the inventory position remains constant during a period, the lump sum charging of holding and shortage costs can be replaced by continuous charging as in the continuous review model.

These observations (cf. Stidham, 1986) enable us to provide a unified treatment of a general class of inventory systems. Assumptions made on the demand process and the cost structure guarantee the optimality of an (s, S) type policy. This and the assumption of full backlogging in turn give rise to a regenerative structure that is exploited throughout the monograph. For completeness, we first establish the fact that the form of the optimal policy is (s, S) by reviewing the classical literature on periodic review systems. The results are then extended to continuous review systems.

1.1 Optimality of (s, S) Policies

The classical multiperiod model (cf. Arrow et al., 1951) is based on the assumption that demands in successive periods are i.i.d. random variables, shortages at the end of any period (except the last) must be made up, and holding and shortage costs are incurred with respect to period ending inventory levels. The one-period expected holding and shortage cost function is assumed convex, and the setup (ordering) cost is independent of the order size. A constant lead time, assumed to be an integral multiple of the period length, is allowed as an extension of the basic model with immediate deliveries. The objective is the minimization of the expected total discounted cost over a planning horizon of n periods. The approach is through dynamic programming based on backward recursion where period n refers to an n-period problem (i.e., n periods left to go).

Although this approach has been the basis for a number of algorithms that are designed to compute optimal inventory control policies, its main utility is in the determination of the form of these policies. For this purpose, we first consider the finite-horizon problem
under immediate deliveries. The discussion is then extended to the
infinite-horizon problem, positive lead times and continuous review
systems.

1.1.1   The Finite-Horizon Problem

For the n-period problem the procurement cost function in every
period is assumed to be of the form:

$$c(z) = \left\{ \begin{array}{ll} K + cz, & z > 0, \\ 0, & z = 0, \end{array} \right. \tag{1.1}$$

where c is the unit cost and K the setup cost. We note in the sequel
that both parameters may be allowed to vary from period to period
(under a monotonicity condition for setup costs) without altering the
main conclusion of this section that (s, S) policies are optimal. The
one-period expected loss function is:

$$\Gamma(y) = \left\{ \begin{array}{ll} \displaystyle\int_0^y h(y-x)b(x)dx + \int_y^\infty p(x-y)b(x)dx & , \quad y \geq 0, \\[3mm] \displaystyle\int_0^\infty p(x-y)b(x)dx & , \quad y < 0, \end{array} \right. \tag{1.2}$$

where y is the quantity on hand after ordering at the beginning of
a period $h(.)$ and $p(.)$ are the inventory holding and shortage cost
functions, respectively, and $b(x)$, $x \geq 0$, is the p.d.f of demand during
a period.

Let $E_n(u)$ denote the total cost of following an optimal policy for
the n-period problem. It is seen that $E_n(u)$ satisfies the functional

equation:

$$E_n(u) = min_{y \geq u} \{c(y-u) + \Gamma(y) + \alpha \int_0^\infty E_{n-1}(y-x)b(x)dx\},$$

$$n = 1, 2, \ldots,$$

(1.3)

where $E_0(u) \equiv 0$, $c(.)$ is the procurement cost function given by (1.1), and $0 \leq \alpha \leq 1$ is the discount factor.

Scarf's original proof of the optimality of (s, S) policies was based on an interesting concept of convexity, that he called $K - convexity$, in relation to the function:

$$G_n(y) = cy + \Gamma(y) + \alpha \int_0^\infty E_{n-1}(y-x)b(x)dx. \qquad (1.4)$$

Definition 1.1 *A differentiable[1] function $g(x)$ is K-convex for $K \geq 0$ if $K + g(a + x) \geq g(x) + ag'(x)$ for $a \geq 0$ and $-\infty < x < \infty$.*

It is seen that 0-convexity is equivalent to ordinary convexity. It also follows from the definition that 1) if g(x) is K-convex then it is M-convex for $M \geq K$, 2) if $g_1(x)$ is K_1-convex and $g_2(x)$ is K_2-convex, then $a_1 g_1(x) + a_2 g_2(x)$ is $(a_1 K_1 + a_2 K_2)$-convex for $a_1 \geq 0$, $a_2 \geq 0$. This last property extends to sums and integrals.

Theorem 1.1 (Scarf, 1960) *If $\Gamma(y)$ is convex then $G_n(y)$ is K-convex.*

Corollary 1.1 *Under the assumption of Theorem 1.1, the optimal policy for the n-period problem is characterized by a pair of critical numbers (s_n, S_n), $s_n < S_n$: if $u > s_n$, order $S_n - u$, if $u > s_n$, do not order where u is the quantity on hand before ordering at the beginning of period n. S_n is the value of y that minimizes $G_n(y)$ and s_n is defined as the smallest number that satisfies $G_n(s_n) = G_n(S_n) + K$.*

---

[1]If differentiability is not assumed, $g'(x)$ is replaced in the definition by $[g(x) - g(x-b)]/b$, $b \geq 0$. In what follows, we assume differentiability but this assumption can be removed (Zabel, 1962).

These follow from Theorem 1.1 because the condition for K-convexity of $G_n(x)$, that is:

$$K + G_n(a + x) \geq G_n(x) + aG_n'(x), \quad a \geq 0, \tag{1.5}$$

rules out the type of shape for $G_n(y)$ that is seen in Figure 1.3a. In that case, we would order up to S_n if $y < s_{n1}$, we would not order if $s_{n1} \leq y \leq s_{n2}$, order if $s_{n2} \leq y \leq s_{n3}$, and would not order if $y > s_{n3}$. Thus the optimal policy would have been specified by four critical numbers. On the other hand, an example for the shape of $G_n(y)$ that is allowed by condition (1.5) is seen in Figure 1.3b. Here, the optimal policy is of (s, S) type as constructed above.

Proof of Theorem 1. If $\Gamma(y)$ is convex, then $G_1(y) = cy + \Gamma(y)$ is convex, therefore K-convex by property 1 we noted above of a K-convex function. Assume $G_n(y)$ is K-convex. It follows from (1.4) and property 3 that $G_{n+1}(y)$ is αK-convex, therefore K-convex if $\int_0^\infty E_n(y - x)b(x)dx$ is K-convex. In turn, again by property 3, this integral is K-convex if $E_n(y)$ is.

Since $G_n(y)$ is K-convex by the induction assumption, the optimal n-th period policy is (s_n, S_n). Therefore:

$$E_n(u) = \begin{cases} K - cu + G_n(S_n) & , u < s_n, \\ \\ -cu + G_n(u) & , u \geq s_n. \end{cases} \tag{1.6}$$

First, for $u \geq s_n$, $E_n(u)$ is K-convex as it is the sum of a linear and a K-convex function. Second, for $a > 0$ arbitrary but fixed and $u < s_n < u + a$, we have:

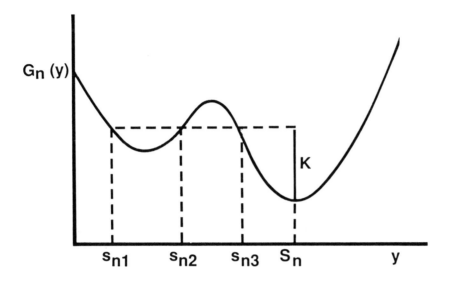

Fig.1.3a. $G_n(y)$ is not K_convex

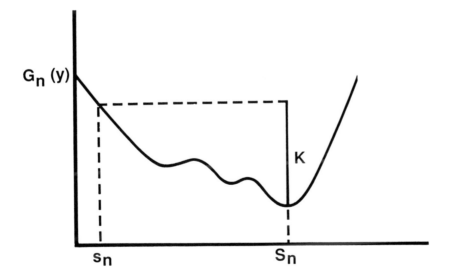

Fig.1.3b. $G_n(y)$ is K_convex

$$E_n(u) \;=\; min_{y \geq u} \left\{ c(y - u) + \Gamma(y) + \alpha \int_0^\infty E_{n-1}(y - x) b(x) dx \right\}$$

$$\leq \; c(a) + \Gamma(u + a) + \alpha \int_0^\infty E_{n-1}(u + a - x) b(x) dx$$

$$= \; c(a) + E_n(u + a). \tag{1.7}$$

The inequality is due to the fact that $y = u + a$ is not necessarily optimal. The third line follows since the optimal policy is not to order when $u + a > s_n$. By definition, $E_n(u)$ is K-convex if $K + E_n(u + a) - E_n(u) - a E_n'(u) = K + E_n(u + a) - E_n(u) + ca \geq 0$ where we used $E_n'(u) = -c$ for $u < s_n$ by (1.6). This holds in view of (1.7) where $c(a) = K + ca$. This shows that $E_n(u)$ is K-convex and completes the proof of the theorem.

Scarf's theorem and proof extends immediately to non-stationary costs and demand distributions. Clearly, dependence on the period of the loss function or the demand distribution (i.e., $\Gamma_n(y)$ instead of $\Gamma(y)$ and $b_n(x)$ instead of $b(x)$) does not alter the proof or the conclusion, provided that $\Gamma_n(y)$ is convex for each n and period demands are independent random variables. Also, since $g(x)$ is K-convex if it is M-convex where $M \leq K$, the set-up costs and discount factors may also vary from period to period so long as $K_n \geq \alpha_n K_{n-1}$.

Aside from these obvious extensions, Veinott (1966) provided an alternative proof of optimality of (s, S) type policies for the n-period inventory problem under different assumptions. He replaced the convexity assumption on the one-period loss function by the weaker assumption of quasi- convexity (i.e., $\Gamma(x)$ is unimodal). More importantly, he extended the class of problems for which there is an optimal (s, S) policy. In addition to complete backlogging, this class includes models in which the excess demand is partially or completely lost (see also Veinott and Wagner, 1965).

1.1.2 The Infinite-Horizon Problem and Stationary Policies

Limiting behavior of the optimal cost functions $E_n(x)$ and policy sequences $\{S_n\}$ and $\{s_n\}$ was first investigated by Iglehart (1963). He showed that $S_1 \leq S_n \leq M < \infty$ and $2s_1 - S_1 < s_n$, $n = 2,3,...,$ provided that $G_1(x)$, assumed convex, attains its minimum at $S_1 < \infty$. (Otherwise the optimal policy is never to order.) Using these bounds, he proved the following result.

Theorem 1.2 (Iglehart, 1963) *$Lim_{n\to\infty} E_n(x) = E(x)$ exists and the convergence is uniform in any finite interval.*

As a corollary to this theorem, it can be seen that the limit function $E(x)$ satisfies the functional equation

$$E(u) = min_{y \geq u}\{c(y - u) + \Gamma(y) + \alpha \int_0^\infty E(y - x)b(x)dx\}$$

Also

$$lim_{n\to\infty} G_n(y) \equiv G(y) = cy + \Gamma(y) + \alpha \int_0^\infty E(y - x)b(x)dx$$

attains its minimum at every limit point S of the sequence $\{S_n\}$.

Since it is the limit of K-convex functions, $G(y)$ is also K-convex, and any limit point s of the sequence $\{s_n\}$ satisfies $G(s) = K + G(S)$ where S is a limit point of the sequence $\{S_n\}$. The optimal policy for the infinite-horizon problem is therefore of (s, S) type.

However, convexity of $\Gamma(y)$ and other assumptions made above for the n-period model are not sufficient to guarantee that $G(x)$ has a unique minimum. This becomes an issue in actual computations of optimal policies as we shall see in Chapter 4.

1.1.3   Positive Lead Times,
Continuous Review Systems

Under the assumptions of the classical model, (s, S) policies remain
optimal in the presence of a constant, positive lead time between
the placement of a replenishment order and its receipt. This delay,
denoted by L, is assumed to be an integral multiple of the period
length. Thus an order placed at the beginning of period n is received
at the beginning of period $n + L$. (In this section, we number the
periods forward.) We assume $n \geq L$.

Using a convention of Beckman (1965), we can charge the system
in period n, as a function of the inventory position, the expected cost
in period $n + L$, discounted for L periods. This is workable because
all orders that are outstanding at the beginning of period n would
have arrived by period $n + L$, and no order placed after period n
would arrive by period $n + L$. Thus the inventory on hand at the
beginning of period $n + L$ is determined by inventory position at the
beginning of period n and demand during the lead time. In view of
this observation, it is seen that the optimality of (s, S) policies, operating on inventory position, extends to the case of constant, positive
lead times.

Optimality of (s, S) policies also extends to the continuous review system introduced at the beginning of the chapter. This system
can be regarded as a periodic review system with random period
lengths $\{X_n, \ n = 1, 2, ...\}$. The inventory position remains the same
throughout a period; it is depleted by batch demands at period ends.
These modifications are inessential, however, to the proof of optimality of (s, S) policies, provided that the sequences $\{X_n\}$ and $\{Y_n\}$ of
period lengths and periodic demands are pairwise independent. Also,
we can replace the discrete charging of holding and shortage costs at
the end of a period with continuous charging during a period as a
function of inventory position. Again the proof of optimality is not
affected provided that the holding and shortage cost rate functions
are convex.

In terms of continuous review systems with constant, positive lead times, Beckman's convention still works. Although the lead time is not now an integral multiple of the length of a period, inventory on hand at time $t + L$, $t \geq L$, will still be equal to inventory position at time t minus the demand during $(t, t + L]$ (see Section 3.2). Therefore, we can again charge the system at time t, as a function of the inventory position, the expected cost rate at time $t + L$.

Finally, we note that the expected discounted cost criterion can be replaced by the expected total cost ($\alpha = 1$) or the expected average cost criterion for the finite-horizon models. In what follows we will be concerned exclusively with the infinite-horizon, stationary model under the long-run average cost criterion.

1.2 Stationary Analysis

Having established the optimality of (s, S) type policies, the next problem is the determination of an optimal (s, S) policy. Traditionally, there have been two main approaches here. The first is based on the dynamic programming formulation of the previous section in a computational setting. This approach is used exclusively for the periodic review version of the model. It has been well-explored since the mid 1960s within the framework of *Markov decision process* (MDP) models. It can be applied to problems with a finite or infinite planning horizon, under one of a wide range of optimization criteria, using stationary or time-varying data. A number of efficient computational algorithms are also available for the solution of these models (cf. Heyman and Sobel, 1984). However, the MDP approach is not routinely applicable in actual practice, except in situations where the setup cost can be ignored (i.e., $c(z) = cz$ in (1.1)). In this case, under a set of mild assumptions, the periodic review model admits a *myopic* optimal policy. This amounts to the reduction of the multiperiod dynamic problem to a forward sequence of one-period, static problems of the classical *newsboy* type. (See Heyman and Sobel, 1984, Chapter 3.)

The second approach has come to be known as *stationary analysis*. It involves the determination of the limiting distribution of on-hand inventory under an (s, S) policy. Using this distribution, an expected cost rate function is constructed with s and S as decision variables. Minimization of this function then determines the optimal (s, S) policy. The approach is applicable to both periodic and continuous review systems. In fact, as pointed out before, the periodic review system can be viewed as a special case of the continuous review system.

1.2.1   Renewal Theory and Inventory Distributions

Stationary analysis of the inventory systems introduced at the beginning of this chapter is greatly helped by renewal theoretic concepts. We review these[2] in chapter 2. Under the assumptions made about the demand process, the sequences $\{X_n, \ n = 1, 2, ...\}$ and $\{Y_n, \ n = 1, 2, ...\}$ representing interarrival times and demand batch sizes form independent renewal processes. We denote by $N(u, u + t)$ the number of arrivals (demands) during $(u, u + t]$, and by $R(t)$ and $r(t)$ the *renewal function* and the *renewal density*, respectively, of the sequence of demand batch sizes. Also, if $D(u, u + t)$ represents the total demand during $(u, u + t]$, $\{D(u, u + t), \ t \geq 0\}$ is a *cumulative renewal process*. Inventory cycles are defined as *first passage times* (of the level $S - s$) of this process. Under the assumption of complete backlogging, they in turn form a renewal process. Thus the inventory position process $\{I_p(t), \ t \geq 0\}$ is *regenerative* (see Figure 1.1). Each time an order is placed, an inventory cycle ends, inventory position is raised to S, and a new cycle of the same probabilistic structure begins.

Time-dependent and limiting distributions of $\{I_p(t), \ t \geq 0\}$ can be determined easily by conditioning on its imbedded renewal process of inventory cycles (section 3.1). For the process $\{I(t), \ t \geq 0\}$ that

---

[2]Chapter 2 provides an informal review of what is needed in renewal theory. It also contains some new material. Readers not familiar with basic renewal theory may wish to refer to Chapter 2 as they read the rest of this chapter.

represents on-hand inventory, the easiest way is to proceed on the basis of the relationship $I(t + L) = I_p(t) - D(t, t + L)$ which holds under the assumption of constant lead times. Note that $D(t, t + L)$ represents the total demand during a lead time that does not necessarily start with an arrival point. (If u is an arrival point, we write $N(u, u + t) \equiv N(t)$ and $D(u, u + t) \equiv D(t)$.) This leads to a complication in that the processes $\{I_p(t), \ t \geq 0\}$ and $\{D(t, t + L), \ t \geq 0\}$ are not, in general, independent. Independence holds in finite time only when the interarrival time distribution is negative exponential. Accordingly, to obtain the distribution of $I(t)$, one needs the joint distribution of $I_p(t)$ and $D(t, t + L)$. This in turn depends on the joint distribution of $N(t)$ and $N(t, t + L)$, representing the numbers of occurrences in the demand process during $(0, t]$ and $(t, t + L]$. This joint distribution is given in section 2.1. (It is extended in section 6.1 to the joint distribution of $N(t)$, $N(t, t + u)$, and $N(t + u, t + u + v)$ for the study of systems with two-shipment order arrivals.)

Based on this background, we determine in section 3.2 the time-dependent and limiting distributions of $\{I_p(t), D(t, t + L)\}$, $I(t)$ and $D(t, t + L)$. One consequence of the results obtained is that $I_p(t)$ and $D(t, t + L)$ are asymptotically independent. Thus if we denote by I_p and $\bar{D}(L)$ the inventory position and demand during a lead time in a stationary process, their marginal distributions can be combined in accordance with $I = I_p - \bar{D}(L)$ to obtain the limiting distribution of on-hand inventory.

The stationary distribution of inventory position turns out to be independent of the interarrival time distribution; it depends only on the renewal function $R(t)$ and the renewal density $r(t)$ of the sequence $\{Y_n, \ n = 1, 2, ...\}$ of batch sizes. This has a simplifying effect on the optimization problems (that we discuss in chapters 4 and 5) in zero-lead-time systems where $I(t) \equiv I_p(t)$. On the other hand, the limiting distribution of the lead time demand depends on the distributions of both interarrival times and demand batch sizes.

The complicated form of the lead time demand distribution is the main cause of the difficulties encountered in the optimization of

inventory systems with positive lead times. To overcome such difficulties, this distribution is sometimes approximated in the literature
by a suitable distribution. In this context, the normal distribution
is often found useful, its relevance being assured by the central limit
theorem when the lead time is reasonably large (i.e., as compared to
the mean interarrival time.) Other distributional forms can be used
to obtain a better approximation than that can be provided by the
normal distribution. However, any such choice should be based on
conceptually firm grounds, not on arbitrary choices.

1.2.2 Derived Measures–The Cost Rate Function

Time-dependent and limiting distributions of inventory position,
lead time demand and on-hand inventory provide a wealth of information about the operating characteristics of the inventory systems
under consideration. They also form the basis for the construction
of a number of other distributions and measures of effectiveness that
arise in relation to certain design and control problems. This wealth
of information is not available through the dynamic programming
approach outlined before.

Of all the different secondary measures that can be defined, we
emphasize in what follows the stationary *fill rate* and the *cost rate
function*. Distribution of the customer waiting time is also discussed
in section 3.3 as a topic of independent interest.

The stationary fill rate, $\bar{F}(0)$, is the probability in steady state
that on-hand inventory is positive. Its complement, $F(0)$, is the
stationary *stockout probability*. Behavior of the fill rate as $\Delta = S - s$
varies will be of considerable interest in the sequel in relation to
optimization problems. We show in section 3.3 that $\bar{F}(0)$ is monotone
increasing in Δ if the expected cycle length (i.e., $1+R(\Delta)$) is log-
concave in Δ. For the continuous review system under immediate
deliveries, this condition turns out to be also necessary. That the
fill rate is monotone increasing in Δ may appear to be self-evident
without any condition. Unfortunately, such is not the case and the

condition that the expected cycle length be log-concave is not a weak one. The same condition also arises as a unimodality condition for the stationary cost rate function (cf. Stidham, 1977, 1986).

The most common measure of effectiveness used in relation to the models under consideration is the stationary cost rate function that is obtained by imposing a cost structure on the system in steady state. In what follows, we choose a simple representation involving a fixed ordering cost and linear holding and shortage costs. The cost rate function then represents the total cost (ordering, holding, and back-logging) per unit time in steady state over an infinite horizon. As we noted before, if we ignore the ordering cost, myopic policies turn out to be optimal over an infinite horizon (or over a finite horizon with an appropriate salvage value) under an average cost or a discounted total cost criterion. Positive ordering (setup) cost is therefore an essential complication. We choose a time-average cost criterion, although a discounted total cost criterion is also possible.

As mentioned, a sufficient condition for the unimodality of the cost rate function is that $1 + R(x)$ be log-concave (Stidham, 1977, 1986). Unfortunately, this condition excludes large classes of demand distributions that may arise in practice. Also, it is not a necessary condition for unimodality except when the lead time is zero in the continuous review system. On the other hand, when this condition holds, the cost rate function is unimodal for any set of values for the ordering, holding, and shortage cost parameters. One suspects that the condition that induces unimodality in this sense should also result in other nice properties. Actually, unimodality of the cost rate function turns out to be equivalent to the property that the fill rate is monotone increasing in Δ, and to the property that the optimal reorder point for a given Δ is monotone decreasing in Δ. All these properties hold if $1 + R(x)$ is log-concave. These facts are proved in Chapter 4.

1.2.2   Exact and Approximately Optimal Policies

The stationary optimal (s, S) policy can be determined in principle by finding the global minimum of the cost rate function. However, computational difficulties make the exact models unattractive in practice. This has motivated several approaches for the determination of approximately optimal (s, S) policies that require less computational effort and demand information. Most of these approaches are still too complicated for practical implementation, however. One exception is the *power approximation* for periodic review systems (Ehrhardt, 1979; Ehrhardt and Mosier, 1984), which is distribution-free and easily computable.

In order to be widely usable in practice, any approximation must fulfill two general requirements: 1) it must be accurate for a pre-specified, wide range of parameter settings, and 2) it must be distribution-free and easily computable. In relation to the first requirement, it should be emphasized that by accurate we mean practically the same as the optimal policy. Under most parameter settings, the cost rate function turns out to be flat around its global minimum. Accordingly, an approximation error of more than one or two percent relative cost increase should be considered excessive. It is important that the range of parameter values for which such accuracy is achieved by an approximation is specified. Clearly, any approximation will perform well under some parameter settings and fail under others.

In view of these considerations, perhaps the most promising approach to developing accurate approximations to stationary control policies that require minimal computational effort and demand information is the one based on asymptotic renewal theory (cf. Roberts, 1962; Ehrhardt, 1979; Ehrhardt and Mosier, 1984; Tijms and Gronevelt, 1984; Sahin and Sinha, 1987). In Chapter 5, we review some of these approximations. Our primary concern in section 5.1 is with the identification of simple, distribution-free conditions under which asymptotic approximations are accurate for continuous and periodic review systems with zero lead times. Under these conditions, we compare the performances of various approximations with those of the optimal policies using gamma, Weibull, truncated normal, inverse Gaussian, and log-normal distributions for demand. In addition to the approximations developed in the monograph, we test in Chapter 5 the accuracy of the *modified power approximation* of Ehrhardt and Mosier (1984) using a wide range of demand distributions and parameter settings.

As noted above, computational problems are more serious for positive lead times due to the complicated form of the lead time demand distribution. Substitution of a more tractable approximating distribution is often accompanied in the literature by other simplifications, involving the measure of effectiveness being used. Most of these are ad-hoc adjustments achieved by droping certain terms that are deemed insignificant. These simplifications are often unnecessary as they do not really simplify the computational problem.

In section 5.2, we discuss the computation and vaidation of optimal policy approximations under positive lead times that are based on: 1) the replacement in the cost rate function of the renewal function $R(x)$ with its linear asymptote, and 2) the normal distribution for the lead time demand distribution as an asymptotic approximation for large lead times. The first represents a continuation of the approach used for the zero lead time model. The second involves the fitting of a normal distribution to the asymptotic mean and variance of the lead time demand distribution.

To validate the various approximations and some of the theoretical findings, it is also necessary to compute the corresponding optimal policies. These computations are based on simple search routines that exploit the sufficient optimality conditions established in Chapter 4. Also required for this process is the accurate numerical computation of the renewal function and related measures. For this purpose, we use the *generalized cubic splining algorithm* of McConalogue (1981) which is basically designed to compute convolutions of a probability distribution. The renewal function is computed from the series representation with an appropriate convergence criterion. Other renewal-theoretic functions and terms that appear in the cost rate function and the optimality conditions are computed by integrating the appropriate spline representations.

1.2.3   Special Cases and Generalizations

Important simplifications can be achieved and sharper results can be obtained when the lead time is zero or when all batch sizes are unity (i.e., the unit-demands model). For the continuous review system with immediate deliveries, it turns out that the fill rate is monotone increasing in Δ, the optimal reorder point for given Δ is monotone decreasing in Δ and the cost rate function is unimodal for all possible values of the cost parameters if and only if $1 + R(x)$ is log-concave (section 4.2.1). Log-concavity of $1 + R(x)$ was first shown to be both necessary and sufficient for the unimodality of the cost rate function in Stidham (1977). This condition is not necessary for periodic review systems and systems with positive lead times. In periodic review systems with zero lead times, unimodality of the cost rate function and the monotonicity of the fill rate and the reorder level prevail also under the condition that the renewal density $r(x)$ is concave increasing (section 4.2.2). Note that if $r(x)$ is monotone decreasing then $R(x)$ is concave and therefore log-concave. However, if $r(x)$ is increasing then $R(x)$ is convex (increasing) and need not be log-concave. Optimization of continuous and periodic-review systems under zero lead times is considered in detail in Chapter 5.

In unit-demand systems, orders can be placed precisely when the inventory position reaches the reorder point. This reduces the (s, S) policy effectively to a *reorder-point, order-quantity* policy. Another feature that is also unique to these systems is the fact that the stationary distribution of inventory position is uniform. Together with a constant procurement lead time, these characteristics imply, as we show in Chapter 4, a unimodal cost rate function without additional assumptions. As a consequence, optimization of the stationary unit-demands model is almost trivial (section 5.1.1). In the case of positive lead times, we approximate in section 5.2.1 the lead time demand distribution by a normal distribution with the same mean and variance as the stationary distribution of lead time demand. The same approximation is also used for the optimization of the general model in section 5.2.2.

In Chapter 6, we consider some extensions of the basic continuous-review model. These include systems where an order may arrive in two shipments, systems with ordering delays, and systems with random lead times. In spite of the fact that an (s, S) policy may no longer be optimal for these systems, our discussion is limited to their operating characteristics under an (s, S) policy. The first two extensions are generalizations of the recent works of Moinzadeh and Lee (1989) and Weiss (1988). For the model with random lead times, we use an approach based on Finch (1961) (cf. Sahin, 1983a) to construct a formal solution for the general model and obtain explicit results in some special cases.

Chapter 2

Renewal Theory Background

A renewal process may be defined as a sequence $\{X_n, n = 1, 2, ...\}$ of independent, identically distributed random variables that are *placed one after another*. We may think of X_n as the interval from the $n-1$st to the n-th occurrences of an *event*, although X_n may have nothing to do with time and may not separate *events* as such. In this chapter, we provide an informal review of the basic concepts and results in renewal theory and introduce the *interval* distributions of interest in the sequel. We also report some new results, both theoretical and empirical, that are required for later use.

2.1   Basic Theory

Consider the sequence $\{ X_n, n = 1, 2, ... \}$ of independent, identically distributed random variables. Let

$$G(x) = P[X_n \leq x], \quad n = 1, 2, ... \ x \geq 0, \tag{2.1}$$

denote the common c.d.f. of the sequence. We assume that $G(x)$
is absolutely continuous with p.d.f. $g(x)$. This assumption is not
restrictive as there is a parallel theory involving discrete representations. We denote by μ and σ^2 the mean and variance of $G(x)$.
Define

$$S_n = X_1 + X_2 + \ldots + X_n, \quad n = 1, 2, \ldots . \tag{2.2}$$

If X_n is interpreted as the time interval separating the $n-1$st and n-th
occurrences of an event, then S_n would be the waiting time until the
n-th occurrence. A renewal process may be defined, alternatively, as
the sequence $\{ S_n, n = 1, 2, \ldots \}$ of partial sums. Clearly, $E[S_n] = n\mu$
and $Var[S_n] = n\sigma^2$. The c.d.f. and p.d.f. of S_n can be expressed in
terms of the convolutions of $G(x)$ and $g(x)$; thus we have

$$G_n(x) = P[S_n \leq x] = \int_0^x G_{n-1}(x - u)g(u)du, \quad n = 2, 3, \ldots, \tag{2.3}$$

$$g_n(x) = \int_0^x g_{n-1}(x - u)g(u)du, \quad n = 2, 3, \ldots, \tag{2.4}$$

with $G_1(x) \equiv G(x)$ and $g_1(x) \equiv g(x)$.

2.1.1   Number of Renewals

Appealing to the *events separated by intervals* interpretation, let
$N(t)$ denote the number of occurrences during $(0, t]$ where the origin coincides with an occurrence that is not counted. Evidently:
$\{N(t) < n\} \iff \{S_n > t\}$. Therefore, $P[N(t) < n] = 1 - G_n(t)$,
and:

$$P[N(t) = n] = G_n(t) - G_{n+1}(t), \quad n = 0, 1, \ldots . \tag{2.5}$$

where $G_0(t) = 1$ if $t \geq 0$ and $G_0(t) = 0$ if $t < 0$. This establishes the
distribution of the number of occurrences over time.

More generally, we may define $N(u, u+t)$ as the number of occurrences during $(u, u + t]$ where $u \geq 0$ is not necessarily an occurrence
point. $N(u, u + t) \equiv N(t)$, if u is an occurrence point. Let

$$w(u, t; n, k) = P[N(u) = n, N(u, u + t) = k],$$
$$u \geq 0, \quad t \geq 0, \quad n, k = 0, 1, \ldots, \tag{2.6}$$

denote the joint distribution of the numbers of occurrences during $(0, u]$ and $(u, u + t]$. We have:

Lemma 2.1

$$w(u, t; n, k) = \begin{cases} 1 - G(u + t), & n = 0, \ k = 0, \\[2mm] \int_0^u g_n(y)[1 - G(u + t - y)]dy, & n \geq 1, \ k = 0, \\[2mm] \int_0^t g(u + y)[G_{k-1}(t - y) - G_k(t - y)]dy, \\ & n = 0, \ k \geq 1, \\[2mm] \int_{x=0}^u g_n(x) \int_{y=0}^t g(u - x + y) \\ \quad \{G_{k-1}(t - y) - G_k(t - y)\}dy \, dx, \\ & n \geq 1, \ k \geq 1. \end{cases}$$

$$(2.7)$$

Proof For $n = k = 0$, we must have the first interval extend beyond $u + t$. For $n \geq 1$, $k = 0$, the n-th occurrence must take place before u (with probability $g_n(y)dy, y < u$) followed by an occurrence-free interval of length at least $u + t - y$. For $n = 0$, $k \geq 1$, we need the first interval terminating after u, say at $u + y$, $y < t$ (with probability $g(u+y)dy$) followed by exactly $k-1$ occurrences during $t-y$. Finally, for $n \geq 1$, $k \geq 1$, the n-th occurrence must take place before time u, with probability $g_n(x)dx$, $x < u$; the following interval should be of length $> u - x$, say $u - x + y$, $y < t$ (this has probability $g(u-x+y)dy$) and, there must occur exactly $k - 1$ events during $t - y$.

The distribution of $N(u, u + t)$ can be obtained from (2.7) as a marginal distribution. If we let

$$k(t) = \sum_{n=1}^{\infty} g_n(t), \quad t \geq 0, \qquad (2.8)$$

we find

$$P[N(u, u+t) = 0] = 1 - G(u+t) + \int_0^u k(y)[1 - G(u+t-y)]dy, \quad (2.9)$$

and, for $k \geq 1$:

$$P[N(u, u+t) = k] = \int_0^t g(u+y)[G_{k-1}(t-y) - G_k(t-y)]dy$$

$$+ \int_{x=0}^u k(x) \int_{y=u}^t g(u-x+y)[G_{k-1}(t-y) - G_k(t-y)]dy \, dx \, .$$

$$(2.10)$$

Note that for $u = 0$, these reduce to (2.5).

2.1.2   Renewal Function and Renewal Density

Let $K(t)$ denote the expected number of occurrences during $(0, t]$. We have, by (2.5):

$$K(t) = \sum_{n=0}^\infty n[G_n(t) - G_{n+1}(t)]$$

$$= \sum_{n=1}^\infty G_n(t). \tag{2.11}$$

$K(t)$ is called the *renewal function*. Its first derivative, given by (2.8), is the *renewal density*. Since $g_n(t)dt$ is the probability that the n-th occurrence takes place during $(t, t+dt]$, $k(t)dt$ can be interpreted as the probability that *an* occurrence takes place during $(t, t+dt]$. This observation helps verify the following integral equation on $k(t)$:

$$k(t) = g(t) + \int_0^t k(u)g(t-u)du. \tag{2.12}$$

For in order to have a renewal at time t, either the first interval must terminate at about t, with probability $g(t)dt$, or there must be a renewal before t, say at u, with probability $k(u)du$, followed by an interval of length $t - u$. On integrating both sides of (2.12), we obtain a similar integral equation on the renewal function $K(t)$; thus:

$$K(t) = G(t) + \int_0^t K(t-u)g(u)du. \tag{2.13}$$

Another measure of interest, the variance of the number of occurences over time, can be expressed in terms of the renewal function and the renewal density as: †

$$Var[N(t)] = 2 \int_0^t K(u)k(t-u)du + K(t) - K^2(t). \qquad (2.14)$$

2.1.3 Recurrence Times

In addition to the renewal density, renewal function and the variance function, it becomes necessary in many applications to characterize the random variables representing the time until the next renewal point and the time since the last renewal point at time t. These are commonly referred to as *backward* and *forward recurrence times*.

Let U_t and V_t denote, respectively, the backward and forward recurrence times at time t. Denote by $u_t(x)$ and $v_t(x)$ their p.d.f.'s and by $U_t(x)$ and $V_t(x)$ their c.d.f.'s. If there is no occurrence during $(0, t]$, then $U_t = t$; thus: $P[U_t = t] = 1 - G(t)$. For $x < t$, the p.d.f. of U_t is given by:

$$u_t(x) = k(t-x)[1-G(x)], \quad x < t. \qquad (2.15)$$

To verify this, note that in order for U_t to be about x, $x < t$, there must be a renewal at about $t - x$, followed by an interval of length at least x. Similarly, the density function of V_t can be written as:

$$v_t(x) = g(t+x) + \int_0^t k(t-u)g(u+x)du. \qquad (2.16)$$

For V_t would be about x if either the first interval terminates at about $t + x$, or there is an occurrence at about $t - u$, followed by an interval of length about $u + x$, $0 < u < t$.

---

† Let $\Pi(z, \theta) = \sum_{n=0}^{\infty} z^n \int_0^{\infty} e^{-\theta t}[G_n(t) - G_{n+1}(t)]dt$, $|z| \le 1$, $Re(\theta) > 0$. We have $\Pi(z, \theta) = [1 - \gamma(\theta)]/[\theta[1 - z\gamma(\theta)]]$, where $\gamma(\theta) = \int_0^{\infty} e^{-\theta t}g(t)dt$. This results in $d^2\Pi(z, \theta)/dz^2|_{z=1} = 2\gamma^2(\theta)/[\theta(1 - \gamma(\theta))^2]$ as the Laplace transform of $E[N^2(t)] - E[N(t)]$. The result follows on inversion and from the definition of variance.

2.1.4 Examples

The above distributions and measures can be expressed in explicit, easily computable forms for exponential and exponential-related interval distributions. These include the mixed-exponential and Erlang distributions and the wider class of *phase-type* distributions (cf. Neuts, 1981, Ch. 2). However, there are computational difficulties in other cases. To compute the renewal function or the renewal density, for example, either an infinite series of iterated convolutions of $G(x)$ or $g(x)$ must be evaluated or an integral equation of the type (2.12) must be solved. We discuss these issues in Section 2 in relation to a number of interval distributions of practical interest. Here, we review some classical examples for illustration.

1. Exponential Intervals

For $g(x) = \lambda e^{-\lambda x}, \quad x \geq 0$, we have:

$$G_n(t) = P[S_n \leq t] = 1 - \sum_{j=0}^{n-1} e^{-\lambda t} \frac{(\lambda t)^j}{j!}. \qquad (2.17)$$

Therefore:

$$P[N(t) = n] = G_n(t) - G_{n+1}(t) = e^{-\lambda t} \frac{(\lambda t)^n}{j!}, \qquad (2.18)$$

and the renewal process reduces to the *Poisson process*. It is easily seen that $P[U_t = t] = e^{-\lambda t}$, $u_t(x) = \lambda e^{-\lambda x}$, $x < t$, and $v_t(x) = \lambda e^{-\lambda x}$, $x \geq 0$. Thus $v_t(x) \equiv g(x)$ for every t, a property enjoyed only by the Poisson process.

Also easily verifiable are the results that $k(t) = \lambda$ and $K(t) = \lambda t$. Thus the renewal density is constant and the renewal function is linear—properties that are also unique to the Poisson process.

2. Erlang Intervals

For $g(x) = \lambda^2 x e^{-\lambda x}, \quad x \geq 0$, we find: $k(x) = \lambda(1 - e^{-2\lambda x})/2$, and $K(x) = (2\lambda x - 1 + e^{-2\lambda x})/4$. Results that are relatively easy to compute can also be obtained for the k-stage Erlang distribution with p.d.f.:

$$g(x) = \lambda e^{-\lambda x}\frac{(\lambda x)^{k-1}}{(k-1)!}, \quad x \geq 0, \tag{2.19}$$

which is a special gamma density with $k \geq 1$ integer. It is easy to see that this is the p.d.f. of the sum of k i.i.d. random variables with a common exponential distribution. A further generality, which again produces easily computable results, is obtained by allowing the parameter λ of the exponential distribution to change. Thus the renewal interval is the sum of k independent *stages,* the i-th having an exponential distribution with parameter λ_i.

3. Mixed-Exponential Intervals

For $g(x) = \pi\lambda_1 e^{-\lambda_1 x} + (1 - \pi)\lambda_2 e^{-\lambda_2 x}, \quad x \geq 0, \quad 0 \leq \pi \leq 1,$ we find:

$$K(x) = \frac{1}{\mu}[x + \frac{C_1}{C_2}(1 - e^{-C_2 x})], \tag{2.20}$$

where $\mu = \pi/\lambda_1 + (1 - \pi)/\lambda_2$, $C_1 = \pi(1 - \pi)(\lambda_1 - \lambda_2)^2/(\lambda_1\lambda_2)$ and $C_2 = \lambda_1\lambda_2\mu$.

More generally, results for all renewal-theoretic measures can be developed without much difficulty for the k-stage mixed-exponential p.d.f.:

$$g(x) = \sum_{i=1}^{k} \pi_i\lambda_i e^{-\lambda_i x}, \quad x \geq 0, \quad \sum_{i=1}^{k} \pi_i = 1. \tag{2.21}$$

If $\pi_i > 0$ then the *stages are in parallel;* with probability π_i, the only stage traversed is stage i, having an exponential distribution with parameter λ_i. If, on the other hand, we chose $\pi_j = \prod_{i\neq j}\lambda_i/(\lambda_i - \lambda_j)$, then (2.21) becomes the p.d.f. of the sum of k independent, exponential random variables, the generalization noted in the previous example with k *stages in series.*

The form (2.21) is thus associated with the well-known *method of stages* employed in the study of non-Markovian processes (cf. Cox and Smith, 1961). It allows for considerable flexibility and computational facility in applications. (For more recent related work, see Whitt, 1982). In turn, the p.d.f. (2.21) is a member of the class of *phase-type distributions*, introduced and widely applied by Neuts (1981), that are again constructed through exponential forms.

2.1.5   Monotonicity Properties

In this section, we note some monotonicity properties of the renewal function and related measures that we shall use in the sequel. Unfortunately, very little can be said about these properties, on general grounds, beyond the fact that $K(x)$ is monotone increasing. However, if $G(x)$ belongs to the class of *decreasing failure rate* (DFR) distributions, then $k(x)$ turns out to be monotone decreasing.

A random variable, X with c.d.f. $G(x)$ is said to have a *DFR distribution* on $x \geq 0$ if the failure rate (or the hazard rate) given by

$$h(x) = \frac{g(x)}{1 - G(x)}, \tag{2.22}$$

is monotone decreasing on $x \geq 0$. *Increasing failure rate* (IFR) distributions are defined in a similar way. Thus $G(x)$ is IFR if $h(x)$ is monotone increasing. It can be seen that the two-stage mixed-exponential distribution of Example 3 above, and, in general, the $k-$stage mixed-exponential distribution having p.d.f. (2.21) with $\pi_i > 0$ are DFR distributions, while the 2-stage Erlang distribution of Example 2 is IFR. In the case of the exponential distribution, $h(x) = \lambda$ satisfies both properties simultaneously. The theory of distributions with monotone failure rates is developed in Barlow et al. (1963) (see also Barlow and Proschan, 1975).

We now state and sketch the proof of a result due to Brown (1980).

Lemma 2.2 *If $G(x)$ is DFR then $k(x)$ is monotone decreasing on $x \geq 0$ and $K(x)$ is concave.*

Proof Through a complicated construction, Brown (1980) shows that if $G(x)$ is DFR, then both U_t and V_t are *stochastically increasing* in t; that is,

$$P[U_t \leq x] > P[U_{t+s} \leq x], \quad s \geq 0, \tag{2.23}$$

and similarity for V_t. On the other hand, we can see by (2.12), (2.15), and (2.22) that

$$k(t) = E[h(U_t)]. \tag{2.24}$$

Now since $h(x)$ is decreasing and U_t is stochastically increasing, it follows that $h(U_t)$ is stochastically decreasing and $k(t)$ is decreasing. This in turn implies that $K(t)$ is concave.

As an example, it can be seen that the renewal density of the mixed-exponential distribution of Example 3 above is decreasing. This is clear in the case of two stages; it also holds for p.d.f. (2.21) with $\pi_i > 0$.

The above result relates the monotonicity of the failure rate function to that of the renewal density for the DFR class. Unfortunately, $G(x)$ IFR does not imply that $k(x)$ is increasing. And little else is known about the extent to which the monotonicity properties of a distribution function are inherited in some way by its renewal process. The monotonicity of the renewal density, when it prevails, is more easily related to the stochastic monotonicity of the forward and backward recurrence times. To this end, we have the following.

Lemma 2.3 U_t and V_t are stochastically decreasing (increasing) in $t \geq 0$ if $k(t)$ is increasing (decreasing) in $t \geq 0$.

Proof We have by (2.15), for $s \geq 0$, $x < t$:

$$P[U_t \leq x] - P[U_{t+s} \leq x] = \int_0^x [1 - G(u)][k(t-u) - k(t+s-u)]du. \tag{2.25}$$

Clearly, this is ≤ 0 (≥ 0) and therefore U_t is stochastically decreasing (increasing) if $k(t)$ is increasing (decreasing). For V_t, we have from (2.16):

$$P[V_t \le x] = K(t+x) - K(t)[1-G(x)] - \int_0^x K(t+x-u)g(u)du, \quad (2.26)$$

and

$$v_t(x) = g(t+x) + \int_0^t k(u)g(t+x-u)du \pm \int_0^{t+x} k(u)g(t+x-u)du$$

$$= k(t+x) - \int_t^{t+x} k(u)g(t+x-u)du. \quad (2.27)$$

From (2.26) we get

$$\frac{d}{dt}P[V_t \le x] = v_t(x) - k(t)[1-G(x)]. \quad (2.28)$$

If k(t) is increasing, then (2.27) implies

$$v_t(x) \ge k(t+x)[1-G(x)] \ge k(t)[1-G(x)], \quad (2.29)$$

and by (2.28) V_t is stochastically decreasing. If $k(t)$ is decreasing, then (2.27) implies

$$v_t(x) \le k(t+x)[1-G(x)] \le k(t)[1-G(x)], \quad (2.30)$$

and by (2.28) V_t is stochastically increasing.

These properties can easily be verified on the examples 2 and 3 above where the renewal densities are monotone.

The last monotonicity result that we shall need in the sequel concerns the ratio $\phi(t) \equiv K(t)/k(t)$. In general, not much can be said about the shape of $\phi(t)$. Below is a sufficient condition for it to be increasing.

Lemma 2.4 $\phi(t)$ *is increasing in $t \ge 0$ if $k(t)$ is decreasing or concave increasing in $t \ge 0$.*

Proof As $K(t)$ is increasing, that $\phi(t)$ is increasing in $t \geq 0$ if $k(t)$ is decreasing is clear. Assume that $k(t)$ is concave increasing. Then, $K(t)$ is convex and $\phi(t)$ is pseudoconvex[†] on $t > 0$. The latter follows easily from definitions. For since $K(t)$ is convex and $k(t)$ is concave, we have for any t_1, t_2:

$$K(t_1) \geq K(t_2) + (t_1 - t_2)k(t_2) \tag{2.31}$$

and

$$k(t_1) \leq k(t_2) + (t_1 - t_2)k'(t_2) \tag{2.32}$$

To show $(t_1 - t_2)\phi'(t_2) \geq 0 \Rightarrow \phi(t_1) \geq \phi(t_2)$, we note that, provided $k(t_2) > 0$, the first inequality means:

$$(t_1 - t_2)[k^2(t_2) - k'(t_2)K(t_2)] \geq 0 \tag{2.33}$$

If we use (2.31) and (2.32) in this, we find $K(t_1)k(t_2) \geq K(t_2)k(t_1)$ or $\phi(t_1) \geq \phi(t_2)$ as required and $\phi(t)$ is pseudoconvex on $t > 0$.

Since $\phi(0) = 0$ and $\phi(t) \geq 0$, $t \geq 0$, $\phi(t)$ cannot decrease on $t > 0$ as this would contradict its pseudoconvexity. This completes the proof.

For the mixed-exponential distribution of Example 3 with any number of stages, $k(t)$ is decreasing and for the 2-stage Erlang distribution of Example 2, $k(t)$ is concave increasing. Therefore, $K(t)/k(t)$ is increasing in both cases.

2.1.6 Cumulative and Delayed Processes

The *cumulative renewal process* is constructed by associating with the n-th interval, X_n, of a renewal process a second random variable, Y_n. The second sequence, $\{Y_n, \ n = 1, 2, ...\}$, is also assumed to be i.i.d.

[†] A real valued differentiable function $\phi(t)$ is said to be pseudoconvex on an open convex set X if $(t_1 - t_2)^t \nabla \phi(t_2) \geq 0$ implies $\phi(t_1) \geq \phi(t_2)$ for any $t_1 \in X$, $t_2 \in X$ (Mangasarian, 1965). Every local minimum of a pseudoconvex function is also global (see, for example, Avriel, 1976, Ch.6).

and may be considered to form a renewal process by itself. In general, the two sequences need not be independent; pairwise independence of $\{X_n, Y_n, \; n = 1, 2, ...\}$ is sufficient for tractability. Also, random variables of the second sequence may be allowed to take any real value. However, we shall restrict ourselves to independent sequences of non-negative valued random variables. We denote by $F(x)$, $x \geq 0$, the common c.d.f. of the second sequence.

We may think of Y_n as the *reward* received at the i-th occurrence of the renewal process $\{X_n, \; n = 1, 2, ...\}$. One quantity of interest is the cumulative reward over time. This can be expressed, with reference to the interval $(0, t]$, as:

$$Z(t) = \sum_{n=1}^{N(t)} Y_n, \;\; t \geq 0. \tag{2.34}$$

The process $\{Z_t, \; t \geq 0\}$ is called a *cumulative renewal process*, a *compound renewal process*, or a *renewal reward process*. The distribution function of $Z(t)$ can be determined through the first principles to be:

$$P[Z(t) \leq x] = \sum_{n=0}^{\infty} P[N(t) = n] P[Z(t) \leq x | N(t) = n]$$

$$= \sum_{n=0}^{\infty} [G_n(t) - G_{n+1}(t)] F_n(x), \;\; t \geq 0, \;\; x \geq 0. \tag{2.35}$$

It can also be seen that $E[Z(t)] = K(t) \, E[Y_n]$ and

$$Var[Z(t)] = K(t) \, Var[Y_n] + (E[Y_n])^2 Var[N(t)]. \tag{2.36}$$

A compound renewal process will be used in the sequel to describe the demand process. Thus if $\{X_n, \; n = 1, 2, ...\}$ represents the inter-demand times (interarrival times of costumers) and $\{Y_n, \; n = 1, 2, ...\}$ the batch sizes demanded, then $Z(t)$ would be the total demand during $(0, t]$. In this context and in other applications, it becomes also necessary to characterize the time until $Z(t)$ exceeds, for the first time, a given quantity. Thus, let $T_w = inf\{t : Z(t) > w\}$. This is the *first passage time* of level w. Clearly, $\{T_w \leq x\} \Leftrightarrow \{Z(x) \geq w\}$. Therefore:

$$P[T_w \leq x] = 1 - \sum_{n=0}^{\infty} [G_n(x) - G_{n+1}(x)]F_n(w), \quad w \geq 0, \ x \geq 0. \quad (2.37)$$

It turns out that

$$E[T_w] = \mu \sum_{n=0}^{\infty} F_n(w) = \mu[1 + Q(w)], \quad (2.38)$$

where $Q(t)$ is the renewal function associated with the sequence $\{Y_n, \ n = 1, 2, ...\}$. Note that in this case the renewal process does not refer to time.

In what follows, we shall also need the joint distribution of T_w and V_w, the latter being the forward recurrence *time* at *time* w, defined with respect to the renewal process $\{Y_n, \ n = 1, 2, ...\}$. We denote the related joint density by:

$$\ell_w(x, y)dx\, dy = P[x < T_w \leq x + dx, \ y < V_w \leq y + dy], \quad x \geq 0, \ y \geq 0. \quad (2.39)$$

It can be seen that:

$$\ell_w(x, y) = \sum_{n=1}^{\infty} g_n(x) \int_{u=0}^{w} f_{n-1}(u)f(w - u + y)du. \quad (2.40)$$

The *delayed renewal process* is constructed by treating the first interval of an *ordinary* renewal process as exceptional. Thus the sequence $\{X_n, \ n = 1, 2, ...\}$ is still mutually independent; however, $P[X_1 \leq x] = \hat{G}(x)$ is different from the common distribution $G(x)$ of the subsequent intervals for $n \geq 2$. Such processes are of both theoretical and applied interest. They arise naturally, for example, if the origin does not coincide with an occurrence.

Allowances can be made easily in the above developments to derive the corresponding expressions for a delayed process. If $*$ denotes convolution, we can see, for example, that:

$$P[S_n \leq x] = \hat{G} * G_{n-1}(x), \quad x \geq 0, \quad n = 1, 2, \ldots \tag{2.41}$$

and

$$K(t) = \sum_{n=0}^{\infty} \hat{G} * G_{n-1}(t), \quad t \geq 0. \tag{2.42}$$

Also, the integral equation (2.12) now takes the form

$$k(t) = \hat{g}(t) + \int_0^t k(u)g(t - u)du, \tag{2.43}$$

and the recurrence time distributions are modified accordingly.

2.1.7   Renewal Equation and Regenerative Processes

Consider the integral equation

$$\phi(t) = q(t) + \int_0^t \phi(t - u)dG(u), \quad t \geq 0, \tag{2.44}$$

where $\phi(t)$ and $q(t)$ are functions that are bounded over finite intervals and $G(t)$ is a distribution function as before. This equation is generally referred to as the *renewal equation*; it arises very frequently in connection with the theory and application of renewal processes. Given $q(t)$ and $G(t)$, we are interested in its solution and the behavior of this solution as $t \to \infty$. Regarding the former, the following is well-known (cf. Feller, 1966, p. 183).

Lemma 2.5 *The renewal equation (2.44) has the unique solution given by:*

$$\phi(t) = q(t) + \int_0^t k(u)q(t - u)du, \tag{2.45}$$

where $k(.)$ is the renewal density induced by $G(.)$.

Proof Denote a convolution by $*$. On substituting (2.45) in (2.44) we find $q * k = q * (g + g * k)$ which holds in view of (2.12). Therefore, (2.45) satisfies (2.44). To show uniqueness, suppose that, in addition to $q + k * q$, $q + r$ is another solution to (2.44). Then the difference $h \equiv r - k * q$ must satisfy $h = g * h$. This implies on iteration that $h = g_n * h$ for all n where g_n is the n-fold convolution of the p.d.f. g. However, since $g_n(x) \to 0$ as $n \to \infty$ for any finite x, $g_n * h \to 0$ as $n \to \infty$. Therefore, $g_n * h = 0$ for all n and $h = 0$.

We shall review some consequences of the renewal equation in terms of the asymptotic behavior of a renewal process in the next section. Here, we first give an application to the determination of the moments of the forward recurrence time distribution. These moments will be required in relation to the random lead time model of Chapter 6. Clearly, they can be computed from the p.d.f. given by (2.16). However, the following result, reported in Coleman (1982), provides a more convenient representation.

Lemma 2.6 *Moments of the forward recurrence time distribution are given by:*

$$E[V_t^n] = E[(X - t)^n] + \int_0^t \{E[(X - u)^n] - (-u)^n\}k(t - u)du,$$

$$t \geq 0, \quad n = 1, 2, \ldots$$

$$(2.46)$$

where $P[X \leq x] = G(x)$.

Proof By conditioning on the time of the first renewal, we can write:

$$E[V_t^n] = \int_0^t E[V_{t-u}^n]g(u)du + \int_t^\infty (u - t)^n g(u)du$$

$$= E[(X - t)^n] + \int_0^t \{E[V_{t-u}^n] - (u - t)^n\}g(u)du \qquad (2.47)$$

This is a renewal equation with $\phi(t) = E[V_t^n] - (-t)^n$ and $q(t) = E[(X - t)^n] - (-t)^n$. Its solution is (2.46).

The renewal equation arises naturally in the study of *regenerative stochastic processes* (Smith, 1955). In many cases, we can identify a renewal process, say $\{T_n, \ \ n = 1, 2, ...\}$, imbedded in a stochastic process of interest. The stochastic process *regenerates* itself at each renewal epoch in the sense that its behavior during T_m for any m, is a probabilistic replica of its behavior during T_1. For example, the process $\{I_p(t), \ \ t \geq 0\}$ of Chapter 1, where $I_p(t)$ is the inventory position at time t, is a regenerative process. Each time an order is placed, the inventory position is raised to S, and the process is *regenerated*. The sequence of i.i.d. intervals between successive orders (i.e., inventory cycles) forms the imbedded renewal process.

Consider a regenerative process, $\{J(t), \ \ t \geq 0\}$, with state space E. Let $G(.)$ be the common c.d.f. of the intervals of its imbedded renewal processes $\{T_n, \ \ n = 1, 2, ...\}$. Define for any $x \in E$:

$$Q(t, x) = P[T_1 > t, \ J(t) \leq x], \ \ t \geq 0, \qquad (2.48)$$

and

$$M(t, x) = P[J(t) \leq x], \ \ t \geq 0. \qquad (2.49)$$

By conditioning on the first renewal epoch of the imbedded renewal process, it is easy to verify that the function M satisfies the renewal equation:

$$M(t, x) = Q(t, x) + \int_0^t g(u) M(t - u, x) du, \qquad (2.50)$$

which by Lemma 2.5 has the unique solution:

$$M(t, x) = Q(t, x) + \int_0^t k(u) Q(t - u, x) du. \qquad (2.51)$$

These representations will be used extensively in the sequel. (For a more formal account of regenerative processes, see, for example, Çinlar, 1975, pp. 293-302.)

2.1.8   Asymptotic Results

As $n \to \infty$, the distribution of the partial sum S_n defined by (2.2) tends to a normal distribution with mean $n\mu$ and variance $n\sigma^2$. This follows from the central limit theorem; thus:

$$lim_{n\to\infty} P[\frac{S_n - n\mu}{\sigma\sqrt{n}} \leq x] = \Phi(x), \qquad (2.52)$$

where $\Phi(x)$ is the standard normal c.d.f. Using the relation $P[N(t) < n] = P[S_n > t]$, it can be seen that $N(t)$ also has an asymptotic normal distribution with mean t/μ and variance $t\sigma^2/\mu^3$.

Of more significance for our purposes are the limiting behavior of the solution of a renewal equation and its implications in terms of asymptotic results. We now state a number of well-known facts that we shall use in the sequel. (For proofs that are omitted, see, for example, Feller, 1966, pp.346-357.) Most of the classical asymptotic results can be obtained from the following lemma that refers to (2.44) and (2.45). (As before, μ and σ^2 are the mean and variance of $G(.)$.)

Lemma 2.7 *If $G(.)$ is non-arithmetic and $q(.)$ is directly Riemann integrable,[†] then*

$$\lim_{t\to\infty} \phi(t) = \frac{1}{\mu} \int_0^\infty q(u)du. \qquad (2.53)$$

There is a paralllel result if $G(.)$ is arithmetic. The condition on $q(.)$ is not restrictive for our purposes. It is not needed if $q(.)$ is a continuous function vanishing outside a finite interval, or if $q(.)$ is monotone and integrable in the ordinary sense. It is introduced to exclude functions that can oscillate widely at infinity, in which case (2.53) may not hold.

When $\sigma^2 < \infty$, we have the following asymptotic result:

[†] For $h > 0$ fixed, denote by $\underline{m}_n$ and $\overline{m}_n$, the min. and max. respectively of $q(t)$ in the interval $(n-1)h \leq t \leq nh$. Define the *upper and lower Riemann Sums* $\overline{\sigma} = h\sum \overline{m}_n$ and $\underline{\sigma} = h\sum \underline{m}_n$. Then, $q(t)$, $t \geq 0$, is directly Riemann integrable if the two sums converge absolutely and if $\overline{\sigma} - \underline{\sigma} < \epsilon$ for h sufficiently small (cf. Feller, 1966, p. 348).

Lemma 2.8 *If $G(.)$ is non-arithmetic with $\sigma^2 < \infty$, then*

$$K(t) = \frac{t}{\mu} + \frac{\sigma^2 - \mu^2}{2\mu^2} + o(1), \quad t \to \infty. \tag{2.54}$$

Proof It is easily verified that the renewal equation (2.58) is satisfied by:

$$q(t) = \frac{1}{\mu} \int_t^\infty [1 - G(u)] du, \tag{2.55}$$

and

$$\phi(t) = 1 + K(t) - \frac{t}{\mu}. \tag{2.56}$$

Therefore, by (2.53), we have

$$lim_{t \to \infty}[1 + K(t) - \frac{t}{\mu}] = \frac{1}{\mu^2} \int_{t=0}^\infty \int_{u=t}^\infty [1 - G(u)] du dt$$

$$= \frac{\sigma^2 + \mu^2}{2\mu^2}, \tag{2.57}$$

from which the Lemma follows.

Lemma 2.8 implies that

$$\lim_{t \to \infty} K(t)/t = \lim_{t \to \infty} k(t) = 1/\mu. \tag{2.58}$$

These results also follow as corollaries to Lemma 2.7 without the assumption of a finite variance.

We shall find the asymptotic approximation to the renewal function provided by (2.54) particularly useful in constructing approximations to stationary optimal control policies in Chapter 5. Higher order asymptotic approximations have also been reported in the literature. Carlsson (1983) shows, for example, that if $G(t)$ is *strongly non-lattice*[†] with finite moments of integer order $m \geq 2$, then:

[†] $G(t)$ is *strongly non-lattice* if $lim_{|\theta| \to 0} inf|1 - \gamma(\theta)| > 0$ where $\gamma(\theta)$ is the characteristic function of $G(t)$.

$$K(t) = \frac{t}{\mu^2} + \frac{\sigma^2 - \mu^2}{2\mu^2} + \frac{S(t)}{\mu^2} + \frac{R * R(t)}{\mu^3} + o(t^{-m} \log t), \quad t \to \infty,$$

$$(2.59)$$

where $R(t) = \int_t^\infty [1 - G(u)]du$, and $S(t) = - \int_t^\infty R(u)du$.

Regarding the recurrence time distributions, on passing to the limits in (2.15) and (2.16), we find that

$$\lim_{t \to \infty} u_t(x) = \lim_{t \to \infty} v_t(x) = \frac{1 - G(x)}{\mu}. \qquad (2.60)$$

Thus, both recurrence time densities have the same limiting form. If we use (2.60) as the p.d.f. of the first interval, an interesting delayed renewal process arises. For $P[X_1 \leq x] = \frac{1}{\mu} \int_0^x [1 - G(u)]du$, and $P[X_n \leq x] = G(x)$, $n \geq 2$, we find that $k(t) = 1/\mu$ and $u_t(x) = v_t(x) = [1 - G(x)]/\mu$. Thus the renewal and recurrence time densities coincide with their limiting forms for $t \geq 0$. Such a renewal process is called an *equilibrium renewal process.*

2.2   Computational Issues

As we noted earlier, the various functions and measures of a renewal process admit easily computable forms only for interval distributions that are certain mixtures of exponentials. In other cases, which include wide classes of distributions of practical interest, the computational effort must be based on numerical methods and approximations. In this section, we consider some distributions that fall into the second group. These distributions will be used in the numerical examples of Chapter 5 to represent demand batch sizes in the continuous review model and periodic demands in the periodic review model.

2.2.1   Interval Distributions

The interval distributions of interest are the *gamma* (GA), *Weibull*
(WL), *truncated normal* (TN), *inverse Gaussian* (IG), and *log-normal*
(LN) distributions. Baxter et al. (1982) have compiled extensive
tables of the renewal function, its integral, and the variance function
for renewal processes based on these distributions. This work as well
as most of our computations in this section and in Chapter 5 are in
turn supported by an algorithm of McConalogue (1981). We outline
this algorithm in the next part. The distributions in question are
all parametrizable by a *shape parameter* and a *scale parameter*. The
p.d.f. of the *gamma distribution* is:

$$g(x) = \frac{1}{\beta^\alpha \Gamma(\alpha)} x^{\alpha-1} e^{-\frac{x}{\beta}}, \quad x > 0, \tag{2.61}$$

where $\alpha > 0$ is the shape parameter and $\beta > 0$ the scale parameter.
Moments of the distribution are:

$$\mu^{(k)} = \beta^k \frac{\Gamma(k+\alpha)}{\Gamma(\alpha)}, \quad k = 1, 2, \dots . \tag{2.62}$$

Note that for integer values of its shape parameter, the gamma distribution reduces to the Erlang distribution of Example 2 above.

The gamma distribution exhibits different characteristics depending on the value of its shape parameter. For $\alpha < 1$, the coefficient
of variation, c, is > 1, the distribution is DFR, and $g(x)$ is monotone
decreasing on $x \geq 0$ with an infinite singularity at the origin. For
$\alpha = 1$, we have $c = 1$ and the distribution reduces to the exponential
distribution. For $\alpha > 1$, we get $c < 1$ and the distribution is IFR.
As $\alpha \geq 1$ increases, c decreases and the shape of $g(x)$ changes from
positively skewed to symmetrical.

Most of these properties are also exhibited by the *Weibull distribution* with p.d.f.:

$$g(x) = \frac{\alpha}{\beta^\alpha} x^{\alpha-1} e^{-(\frac{x}{\beta})^\alpha}, \quad x > 0, \tag{2.63}$$

and moments

$$\mu^{(k)} = \beta^k \Gamma(1 + \frac{k}{\alpha}), \quad k = 1, 2, \dots . \qquad (2.64)$$

Again, $\alpha > 0$ is a shape parameter and β a scale parameter. Of the properties mentioned above for the gamma distribution, the only difference in the case of the Weibull is that as α increases $g(x)$ changes from positively skewed to negatively skewed.

The *truncated normal* distribution with parameters α and β has the p.d.f.:

$$g(x) = \frac{1}{\beta\sqrt{2\pi}[1 - \Phi(-\frac{\alpha}{\beta})]} e^{-\frac{1}{2}(\frac{x-\alpha}{\beta})^2}, \quad x \geq 0, \qquad (2.65)$$

where $\Phi(x)$ is the c.d.f. of the standard normal distribution. Note that if X has a TN distribution with parameters α and β, then $Y = X/\beta$ has a TN distribution with parameters α/β and 1. This means that β is a scale parameter of the distribution; α is a shape parameter. The moments are given by:

$$\mu^{(k)} = \beta^k k! \frac{I_k(-\alpha/\beta)}{I_0(-\alpha/\beta)}, \quad k = 1, 2, \dots , \qquad (2.66)$$

where

$$I_k(x) = \frac{1}{\sqrt{2\pi}} \int_x^\infty \frac{(u - x)^k}{k!} e^{-\frac{u^2}{2}} du. \qquad (2.67)$$

As in the case of the GM and WL distributions, the TN p.d.f. is also unimodal, and, like the GM density, it is positively skewed. As the shape parameter α increases, c decreases and $g(x)$ becomes more and more *peaked*, but it remains positively skewed.

Another unimodal and positively skewed p.d.f. is provided by the *inverse Gaussian* distribution where:

$$g(x) = \sqrt{\frac{\lambda}{2\pi x^3}} e^{-\frac{\lambda}{2x}(\frac{x-\mu}{\mu})^2}, \quad x > 0. \qquad (2.68)$$

The mean and variance are μ and μ^3/λ. If we make the transformation, $\alpha = \lambda/\mu$, $\beta = \mu^2/\lambda$ and $Y = X/\beta$, then

$$g(x) = \sqrt{\frac{\alpha^2}{2\pi x^3}}e^{-\left(\frac{x-\alpha}{2x}\right)^2}, \quad x > 0. \tag{2.69}$$

Thus, $\alpha > 0$ is a shape parameter and β is a scale parameter.

The last member of our set is the *lognormal distribution* with p.d.f:

$$g(x) = \frac{1}{x\rho\sqrt{2\pi}}e^{-\frac{1}{2}\left(\frac{\log x - h}{\rho}\right)^2}, \quad x > 0, \tag{2.70}$$

and moments:

$$\mu^{(k)} = e^{k\lambda + \frac{1}{2}k^2\rho^2}, \quad k = 1, 2, \dots . \tag{2.71}$$

If we put $\beta = e^\lambda$ and $Y = X/\beta$, then

$$g(x) = \frac{1}{x\rho\sqrt{2\pi}}e^{-\frac{1}{2}\left(\frac{\log x}{\rho}\right)^2}, \quad x > 0. \tag{2.72}$$

Therefore β is a scale parameter and $\rho^2 \equiv \alpha$ is a shape parameter. The density function is very close to the normal density for small α. As α increases so does c and the skeweness increases rapidly.

Thus, all five distributions are parametrizable with a shape parameter and a scale parameter, generally denoted by α and β, respectively. In what follows, we shall carry out the computations for $\beta=1$. The corresponding results for other scale parameter values will be obtainable by simple transformations. Using the scale parameter β as the first argument in notation, we have by definition: $G(\beta; x) = G(1; x/\beta)$, $x \geq 0$, and $g(\beta; x) = g(1; x/\beta)/\beta$. It follows that $\mu(\beta) = \beta\mu(1)$, $\sigma^2(\beta) = \beta^2\sigma^2(1)$ and $c(\beta) = c(1)$. Thus the coefficient of variation is invariant under the scale parameter.

For convolutions of $G(w; x)$, we have:

$$G_2(\beta; x) = \int_0^x G(\beta; x - u)g(\beta; u)du$$

$$= \int_0^{x/\beta} G(1; x/\beta - u)g(1; u)du = G_2(1; x/\beta). \tag{2.73}$$

It follows by induction that $G_k(\beta; x) = G_k(1; x/\beta)$, $k = 1, 2, \ldots$. Consequently, for the renewal function and related quantities, we have: $K(\beta; x) = K(1; x/\beta)$, $x \geq 0$, $k(\beta; x) = k(1; x/\beta)/\beta$, $x \geq 0$, $v_t(\beta; x) = v_{t/\beta}(1; x/\beta)/\beta$, $t \geq 0$, $x \geq 0$, and $u_t(\beta; x) = u_{t/\beta}(1; x/\beta)$ $/\beta$, $t \geq 0$, $x \geq 0$. Similar relations for other quantities can also be developed from these.

2.2.2 Shape of the Renewal Function

For distributions considered above, computation of the renewal function and related quantities must be based on numerical methods. We used the *generalized cubic splining algorithm* of McConalogue (1981). As its predecessor, due to Cleroux and McConalogue (1976), this algorithm provides a general approach to the numerical evaluation of the various renewal-theoretic functions and measures. The essence of the algorithm is the approximation of the n-fold convolution $G_n(t)$ by a cubic-spline function. The interval $[0,t]$ is divided into m panels of equal width, and $G_n(t)$ is approximated by a different cubic function in each panel. The approximation for $G_{n+1}(t)$ is obtained, recursively, from $G_n(t)$. (For details, see Cleroux and McConalogue, 1976; and McConalogue, 1978.)

The generalized algorithm allows $g(t)$ to be unbounded near the origin (but assumes that $g_n(t)$ is bounded for $n \geq 2$). This permits the treatment of both the gamma and the Weibull distributions for shape parameter values less than 1. The algorithm also yields better accuracy near the origin when $g(t)$ is bounded. Using this algorithm, Baxter et al. (1982a,b) computed and tabulated the renewal function, its integral, and the variance function for renewal processes generated by the above distributions. They computed the renewal function through (2.11) with appropriate convergence criteria on $G_n(t)$, and $K(t)$ relative to its linear asymptote. The variance function was computed from (2.14) and the integral of $K(t)$ by directly integrating the spline representation. For the renewal function, they report an accuracy level of four to six decimal figures. Although Baxter et al. tables are very extensive, we were not able to use them directly for

most of our work (cf. Chapter 5). However, in the treatment of the
distributions selected, and in the implementation of the algorithm, we
followed the same approaches, procedures, and convergence criteria
as in the above references.

It is clear that the properties of a distribution (moment characteristics, shape, aging, etc.) will be reflected in some fashion by
its renewal process. However, the underlying relationships are not
well-understood except, as noted earlier, in the case of the DFR class
and its close relatives. Of particular interest to us in what follows
is the shape of the renewal function and the rapidity with which it
approaches its linear asymptote. Unfortunately, given a distribution,
there is no general analytic means of characterizing the shape of its
renewal function. Conversely, it is not clear as to what class of distributions are implicated by a given assumption on the shape of $K(x)$
or $k(x)$. The same is true of the rate of approach of $K(x)$ to its
asymptote.

In general, in approaching its limit $1/\mu$, the renewal density $k(x)$
may be monotone (increasing or decreasing) or it may oscillate. In the
first case the renewal function would be convex or concave increasing
while in the second case it would oscillate around its linear asymptote.
The latter would mean that $K(t) - t/\mu$ may not be monotone, not a
favorable observation for the accuracy of asymptotic approximations
through (2.54). On general grounds, oscillations may arise when $G(x)$
has small dispersion (i.e. $c < 1$). For then $k(t) = \sum_1^\infty g_n(t)$ will tend
to be larger near $t = \mu, 2\mu, \ldots$ and smaller near $t = 0, 1.5\mu, \ldots$ (cf.
Cox, 1962).

Computed through McConalogue's algorithm, shapes of the renewal function typical of the distributions introduced above are seen
in Figures 2.1 to 2.5. In all cases the scale parameter β was taken
as unity and the shape parameter α was varied to generate different
shapes and rates of approach to the linear approximation. As anticipated by Lemma 2.2, the renewal function is concave for the DFR
gamma and Weibull distributions ($\alpha < 1$). We found that $K(t)$ is
nearly concave, save for a possibly convex shape very near the origin,

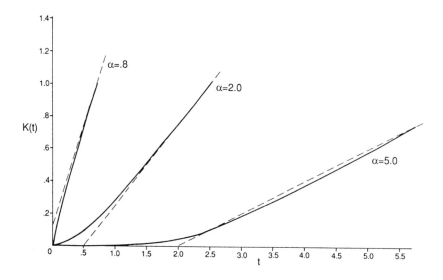

Fig.2.1. Renewal function of the gamma distribution

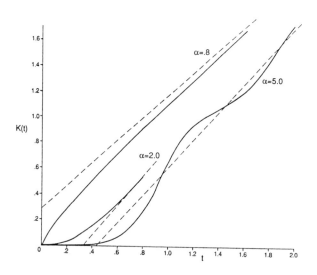

Fig.2.2. Renewal function of the Weibull distribution

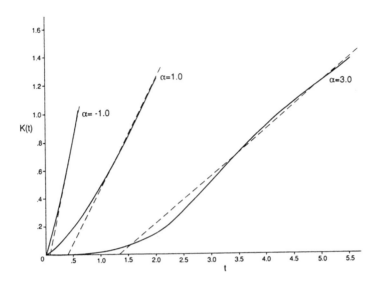

Fig.2.3. Renewal function of the truncated normal distribution

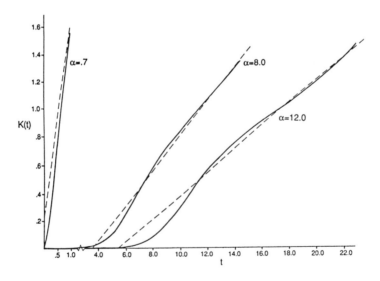

Fig.2.4. Renewal function of the inverse Gaussian distribution

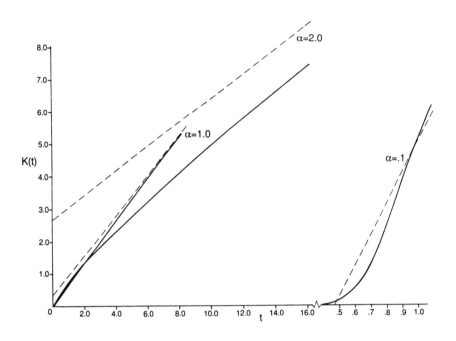

Fig.2.5. Renewal function of the lognormal distribution

also for the inverse Gaussian distribution with $\alpha < 1$. For these three distributions, as $\alpha > 1$ is increased, $K(t)$ becomes initially convex and then oscillatory.

For the truncated normal distribution, as α increases (from -2), the renewal function changes from a convex to an oscillatory shape. For the lognormal distribution, $K(t)$ is oscillatory, following a convex start, for small values of the shape parameter. Oscillations disappear by about $\alpha = 0.8$. For $\alpha \geq 1$, $K(t)$ is concave, with extremely slow convergence to the linear asymptote when $\alpha > 1.5$.

2.2.3   Accuracy of Asymptotic Approximations

The above observations on the shape of the renewal function indicate that the accuracy of the asymptotic approximation is contingent on a number of factors. Important among these are the presence of oscillations (small dispersion) and a low rate of approach to the linear asymptote (high dispersion). From a practical point of view, it would be very useful if these could be made more precise so that one knows how large a t is large enough for sufficient accuracy. To this end, we now let:

$$J(t) = \frac{|K(t) - [t/\mu + (c^2 - 1)/2]|}{K(t)} \tag{2.74}$$

and define:

$$\rho(\epsilon) = inf\{\rho : \; J(\rho\mu) = \epsilon, \;\; J(t) \leq \epsilon, \;\; t \geq \rho\mu\}, \tag{2.75}$$

where the relative error $\epsilon > 0$ is arbitrary but fixed. By a large enough t, we now mean $t \geq \rho(\epsilon)\mu$. This definition is in view of the fact that $|K(t) - t/\mu|$ may not be monotone decreasing as we have seen in the above examples. If $|K(t) - t/\mu|$ is monotone decreasing, then $\rho(\epsilon)$ can be defined simply by $J(\rho(\epsilon)\mu) = \epsilon$.

Using McConalogue's algorithm, we computed the convergence measure $\rho(\epsilon)$ for all the distributions discussed above. The results are presented in Table 2.1 for $\epsilon = 0.01, 0.05$, and 0.10 in columns headed by ρ. It is seen that, generally speaking, $\rho(\epsilon)$ decreases as c increases to about 0.8 and then it increases with c. Increasing $\rho(\epsilon)$ with increasing $c > 1$ can be explained by the higher order asymptotic result noted above (cf. 2.59). The third and fourth terms in that expression imply that the rate of convergence of $K(t)$ would be affected by the shape of $g(t)$, the location of its mode and the prominence of its upper tail. These influences are also reflected in the coefficient of variation. On the other hand, increasing $\rho(\epsilon)$ with decreasing $c < 1$ is due to oscillations of $K(t)$.

Motivated by these observations, we tried a polynomial approximation to $\rho(\epsilon)$ through c. A second degree polynomial of the form;

$$\hat{\rho}(\epsilon) = a_0(\epsilon) + a_1(\epsilon)c + a_2(\epsilon)c^2 \tag{2.76}$$

Table 2.1 Convergence Measures

	α	μ	c	$\epsilon = 0.01$			$\epsilon = 0.05$			$\epsilon = 0.10$		
				ρ	$\hat{\rho}_1$	$\hat{\rho}_2$	ρ	$\hat{\rho}_1$	$\hat{\rho}_2$	ρ	$\hat{\rho}_1$	$\hat{\rho}_2$
GM	7.0	7.00	0.38	1.35	1.23	1.60	0.91	0.87	1.04	0.83	0.78	0.79
	4.0	4.00	0.50	1.01	1.00	1.13	0.54	0.71	0.82	0.52	0.63	0.65
	3.0	3.00	0.58	0.71	0.90	0.98	0.60	0.63	0.73	0.54	0.55	0.59
	2.0	2.00	0.71	0.91	0.81	0.98	0.63	0.54	0.70	0.53	0.46	0.57
	1.5	1.50	0.82	0.95	0.81	1.24	0.58	0.51	0.77	0.46	0.41	0.61
	0.6	0.60	1.29	1.70	1.65		0.83	0.78		0.57	0.53	
WL	7.0	0.94	0.17	3.94	3.30	3.06	2.32	2.00	1.71	1.39	1.33	1.22
	4.0	0.91	0.28	2.07	2.51	2.19	1.42	1.55	1.71	0.99	1.07	0.97
	3.0	0.89	0.36	1.64	2.02	1.70	1.04	1.27	1.09	0.93	0.91	0.82
	2.0	0.89	0.52	1.24	1.31	1.08	0.65	0.85	0.79	0.60	0.68	0.63
	1.5	0.90	0.68	0.99	0.91	0.95	0.73	0.58	0.69	0.62	0.53	0.57
	0.6	1.50	1.76	5.86	6.00		2.76	2.84		1.84	1.84	
TN	4.0	4.00	0.25	2.45	2.40	2.40	1.43	1.42	1.41	0.97	0.91	1.03
	2.0	2.06	0.46	1.27	1.48	1.26	1.04	1.02	0.88	0.57	0.73	0.69
	1.0	1.29	0.62	1.38	1.16	0.95	0.74	0.84	0.70	0.66	0.65	0.58
	0.0	0.80	0.76	1.16	1.15	1.07	0.84	0.79	0.72	0.69	0.61	0.58
	-1.0	0.53	0.85	1.33	1.29	1.35	0.83	0.80	0.81	0.62	0.61	0.64
	-2.0	0.37	0.91	1.34	1.43	1.62	0.75	0.83	0.91	0.52	0.62	0.69
IG	20.0	20.00	0.22	2.73	2.29	2.64	1.73	1.46	1.52	1.27	1.09	1.10
	12.0	12.00	0.29	1.82	1.95	2.12	1.26	1.26	1.28	0.89	0.94	0.95
	8.0	8.00	0.35	1.34	1.67	1.75	0.87	1.09	1.11	0.83	0.81	0.84
	4.0	4.00	0.50	0.84	1.23	1.13	0.70	0.81	0.82	0.60	0.60	0.65
	1.0	1.00	1.00	2.16	1.77	2.15	1.14	0.95	1.11	0.82	0.68	0.82
	0.5	0.50	1.41	4.32	4.62	6.52	2.21	2.36	2.82	1.55	1.65	1.88
LN	0.1	1.05	0.32	1.39	1.42	1.93	0.91	0.88	1.19	0.86	0.82	0.89
	0.5	1.28	0.81	1.29	1.88	1.21	0.94	1.03	0.76	0.68	0.80	0.61
	1.0	1.65	1.31	6.13	6.12	5.16	2.73	2.69	2.28	1.82	1.78	1.54
	1.5	2.12	1.87	15.19	15.22	15.23	6.28	6.29	6.32	4.04	4.04	4.06
	2.0	2.72	2.53		32.21	34.20		13.05	14.27	8.27	8.39	9.03

Table 2.2 Approximating Polynomials

Dist.	$\epsilon = 0.01$			$\epsilon = 0.05$			$\epsilon = 0.10$		
	a_0	a_1	a_2	a_0	a_1	a_2	a_0	a_1	a_2
GM	2.510	-4.150	2.976	1.640	-2.619	1.512	1.467	-2.619	1.512
WL	4.757	-9.680	5.908	2.810	-5.350	3.054	1.286	-3.032	1.742
TN	4.236	-9.011	6.523	2.178	-3.611	2.340	1.239	-1.551	0.957
IG	3.847	-8.395	6.320	2.372	-4.840	3.415	1.745	-3.493	2.425
LN	3.085	-7.581	7.547	1.566	-3.095	3.017	1.360	-2.321	2.015
All[†]	4.830	-12.195	9.486	2.500	-5.345	3.951	1.720	-3.383	2.479

[†]Except DFR GM and DFR WL.

turned out to provide satisfactory fits. Convergence measures estimated by individual fits (separately for each distribution) are given in Table 2.1 in columns headed by $\hat{\rho}_1$

We also tried several combined fits. Examination of the plots of $\rho(\epsilon)$ showed a similar shape for all distributions for about $c < 0.9$. However, three different shapes were observed for $c > 0.9$. In this range, convergence measures for LN were considerably larger in value than those for GM and WL; results for IG and TN fell somewhere in between. Estimates based on grouping GM and WL and TN and IG were similar to $\hat{\rho}_1$.

As GM and WL distributions with $\alpha < 1$ are somewhat exceptional, having densities with infinite singularities at the origin, we also tried a combined fit after deleting these. The remaining distributions, including GM and WL with $\alpha > 1$, have unimodal densities with modes at $t > 0$. Results of this combined fit are also shown in Table 2.1 under column headings $\hat{\rho}_2$. They again show good agreement with the exact results. Coefficients of the approximating polynomials are given in Table 2.2. It is interesting to note that, except for the LN distribution, $K(t)$ can be approximated by its linear asymptote with a small relative error even for small to moderate values of t, as measured in units of μ. It is also interesting that most of the variation in ρ can be explained by c. In most applications, ϵ need not be too small. In the applications reported in Chapter 5, sufficient accuracy is achieved for $\epsilon = 0.20$ which results in even smaller ρ values.

Table 2.2 is based on a larger set of observations than those seen in Table 2.1. However, a restriction was $c < 2$. This is due to the fact that larger coefficients of variation for which $\rho(\epsilon)$ is computable are produced only by the LN distribution. McConalague's algorithm is unable to cope with GM and WL distributions with $\alpha < 0.5$ for which $c > 2$ also.

We emphasize that all the distributions considered are parametrizable in terms of a scale and a shape parameter. Therefore, both c and $\rho(\epsilon)$ depend only on the shape parameter. In terms of scaling, note that $J(\beta; t) = J(1; t/\beta)$, $J(\beta; \rho\mu(\beta)) = J(1; \rho\mu(1))$, and $\rho(w; \epsilon) = \rho(1; \epsilon)$. Thus, in addition to c, ρ is also invariant under the scale parameter. Furthermore, with the elimination of the DFR gamma and DFR Weibull distributions, all the densities have the same unimodal shape with modes at $x > 0$. Within this class, the approximation ρ_2 should be useful in practice by providing an immediate indication of whether the asymptotic approximation would be sufficiently accurate. Outside this class, however, the approximations suggested may be totally inadequate. For example, the convergence measure for a renewal process governed by a mixed-exponential distribution as in (2.21) cannot be well-approximated by c alone. This distribution cannot be parametrized as above. In fact, it admits different parameter configurations under which the mean and variance remain the same but the measure of convergence vary substantially.

Chapter 3

Operating Characteristics

This chapter is concerned with the time-dependent and stationary behavior of the inventory systems introduced in Chapter 1. Details are given for the continuous review system under a compound renewal demand process. The periodic review system is treated as a special case, following the correspondence explained in Chapter 1. Included in the analysis are the stochastic processes representing inventory position, inventory on hand, and demand during a lead time. The investigation is based on the joint distribution of inventory position and demand during a lead time. Time dependent and stationary distributions of lead time demand, inventory position and inventory on hand are derived in section 3.2 from this joint distribution. Section 3.3 is concerned with certain stationary measures of interest that can be constructed from the distributions obtained before. Important among these are the probability that on hand inventory is positive (the fill rate), the cost rate function, and the distribution of the customer waiting time. The chapter is concluded with a discussion of two special cases. This includes an independent treatment of the unit demands model and the specialization of some of the results to the periodic review mode.

For the general model under continuous review, we denote by

$A(x)$, $x \geq 0$, the interdemand time c.d.f with mean μ_a and by $B(x)$, $x \geq 0$, the batch size c.d.f. with mean μ_b, variance σ_b^2 and coefficient of variation c_b. Both distributions are regarded as absolutely continuous for convenience with p.d.f.'s $a(x)$ and $b(x)$, respectively. Inventory position and inventory on hand at time t are denoted by $I_p(t)$ and $I(t)$, respectively. $L \geq 0$ is the length of the constant lead time, $S \geq 0$ is the order-up-to level, and, s, which is allowed to be negative, is the reorder point. We write $\Delta = S - s$ and assume $\Delta \geq 0$ for a meaningful control policy.

3.1   Inventory Position

According to our assumptions on the demand process, the sequence of interarrival times forms a renewal process. We denote by $N(\tau, \tau + u)$ the number of occurrences in this process during $(\tau, \tau + u]$. If τ is an arrival point, then $N(\tau, \tau + u) \equiv N(u)$ and

$$P[N(u) = n] = A_n(u) - A_{n+1}(u), \quad n = 0, 1, \dots . \tag{3.1}$$

A second renewal process may be constructed through the i.i.d. sequence of batch sizes. We let

$$R(x) = \sum_1^\infty B_n(x) \tag{3.2}$$

and

$$r(x) = \sum_1^\infty b_n(x) \tag{3.3}$$

denote the related renewal function and renewal density. Note that this process does not refer to time. $R(x)$, for example, represents the expected number of customers whose demand can be fully satisfied with quantity x.

Let $D(\tau, \tau + u)$ represent the cumulative demand during $(\tau, \tau + u]$. $\{D(\tau, \tau + u), \ u \geq 0\}$, $\tau \geq 0$, is a delayed cumulative renewal process (cf. Section 2.1.6). If τ is an arrival point, then $D(\tau, \tau + u) \equiv D(u)$ and

$$P[D(u) \leq x] = \sum_{n=0}^{\infty} [A_n(u) - A_{n+1}(u)]B_n(x), \quad u \geq 0, \ x \geq 0. \quad (3.4)$$

If we start observing the inventory position immediately after an order $(t = 0)$, we have a well-defined stochastic process $\{I_p(t), \ t \geq 0\}$ with $I_p(0) = S$ (Figure 1.1). Instants at which the inventory position is raised to S are the regeneration points of this process. The length of time between two successive such points is an inventory cycle. Let

$$f_p(t, x)dx = P[x < I_p(t) \leq x + dx], \quad t \geq 0, \ s \leq x \leq S, \quad (3.5)$$

and

$$f_p(x) = \lim_{t \to \infty} f_p(t, x), \quad x \geq 0. \quad (3.6)$$

Also, denote by I_p the inventory position in a stationary process. We have the following result.

Theorem 3.1

$$P[I_p(t) = S] = 1 - A(t) + \int_0^t m(t - u)[1 - A(u)]du, \quad (3.7)$$

$$f_p(t, x) = \sum_{n=1}^{\infty} [A_n(t) - A_{n+1}(t)]b_n(S - x)$$

$$+ \sum_{n=1}^{\infty} b_n(S - x) \int_0^t m(t - u)[A_n(u) - A_{n+1}(u)]du, \quad s \leq S,$$

$$(3.8)$$

where $m(x)$ is the solution of the integral equation:

$$m(x) = c(x) + \int_0^x m(x - u)c(u)du, \quad (3.9)$$

with

$$c(x) = \sum_{k=0}^{\infty} [a_{k+1}(x) - a_k(x)]B_k(S - s). \quad (3.10)$$

Proof If T_n denotes the length of the n-th cycle, then $\{T_n, \quad n = 1, 2, ...\}$ is a renewal process. Denote by $C(x)$, $c(x)$, and $m(x)$, respectively, the c.d.f. and p.d.f. of the cycle lengths and the related renewal density. We see that $\{T_n > x\}$ if and only if $\{D(x) < \Delta\}$, $\Delta \equiv S - s$. It follows from (3.4) that

$$C(x) = 1 - \sum_{n=0}^{\infty} [A_n(x) - A_{n+1}(x)] B_n(\Delta), \quad x \geq 0. \qquad (3.11)$$

The corresponding p.d.f. is (3.10) and the renewal density is determined by (3.9).

The theorem now follows by conditioning on the renewal process of cycles. For (3.7), by conditioning on T_1, we can write:

$$P[I_p(t) = S] = 1 - A(t) + \int_0^t c(u) P[I_p(t - u) = S] du. \qquad (3.12)$$

This is a renewal equation with solution (3.7) (cf. Lemma 2.5). The expression for $f_p(t, x)$ follows from a similar argument.

Corollary 3.1 *For the limiting distribution of inventory position, we have:*

$$P[I_p = S] = \frac{1}{1 + R(\Delta)}, \qquad (3.13)$$

$$f_p(x) = \frac{r(S - x)}{1 + R(\Delta)}, \quad s \leq S. \qquad (3.14)$$

Proof The expected cycle length is given by:

$$E[T_n] = \int_0^{\infty} [1 - C(x)] dx = \mu_a[1 + R(\Delta)]. \qquad (3.15)$$

It follows, by (2.69), that

$$\lim_{t \to \infty} m(t) = \{\mu_a[1 + R(\Delta)]\}^{-1}. \qquad (3.16)$$

If we pass to limits as $t \to \infty$ in (3.7) and (3.8), we obtain (3.13) and (3.14), on account of Lemma 2.6.

Note that the interarrival time distribution has no effect on the limiting distribution of inventory position. Note also that the more natural initial condition, $I_p(0) = y$, $s < y < S$, introduces only minor complications in the above developments. $\{T_n, \quad n = 1, 2, ...\}$ would then define a delayed renewal process with the distribution of T_1 being different ($y - s$ instead of Δ in (3.6)). If $\bar{m}(x)$ is the renewal density for this delayed process, then we have:

$$P[I_p = y] = 1 - A(t), \tag{3.17}$$

$$P[I_p(t) = S] = \int_0^t \bar{m}(t - u)[1 - A(u)]du, \tag{3.18}$$

and

$$f_p(t, x) = \begin{cases} \sum_{n=1}^{\infty}[A_n(t) - A_{n+1}(t)]b_n(y - x) \\ \quad + \sum_{n=1}^{\infty} b_n(S - x) \int_0^t \bar{m}(t - u)[A_n(u) - A_{n+1}(u)]du, \\ \hspace{6cm} s \leq x < y, \\ \\ \sum_{n=1}^{\infty} b_n(S - x) \int_0^t \bar{m}(t - u)[A_n(u) - A_{n+1}(u)]du, \\ \hspace{6cm} y \leq x < S. \end{cases} \tag{3.19}$$

On passing to limits as $t \to \infty$ we obtain the same limiting distribution (3.13), (3.14).

3.2 Inventory On Hand and Leadtime Demand

To discuss the distribution of inventory on hand in finite time, we take $I_p(0) = S$, immediately after a replenishment order, and concentrate on the characterization of $I(t + L)$ for $t \geq 0$. First, let $O(t)$ denote total outstanding orders at time t and $M(t, t + L)$ the total addition to inventory during $(t, t + L]$. We have $I(t) = I_p(t) - O(t)$ and

$I(t+L) = I(t) + M(t, t+L) - D(t, t+L)$. $O(t)$ must be converted to inventory over $(t, t + L]$; that is, $O(t) = M(t, t + L)$. These relations imply:

$$I(t + L) = I_p(t) - D(t, t + L). \tag{3.20}$$

In view of this observation, $I(0)$ is irrelevant for $I(t + L)$, $t \geq 0$. It would be needed only to determine $I(u)$, $0 \leq u < L$. For this purpose, timing and magnitudes of outstanding orders at time 0 would have to be specified also, as part of the initial conditions. For example, if there is only one outstanding order of magnitude y at time 0, due to arrive at time $\tau < L$, and if $I(0) = z$, $z + y = S$, then $I(u) = z - D(u)$ for $u < \tau$ and $I(u) = S - D(u)$ for $\tau \leq u < L$. It follows that $P[I(u) \leq x] = 1 - P[D(u) \leq z - x]$, $x \leq z$, $u < \tau$, and $P[I(u) \leq x] = 1 - P[D(u) \leq S - x]$, $x \leq S$, $\tau \leq u < L$.

To determine the distribution function of $I(t + L)$, $t \geq 0$, we need the joint distribution of $I_p(t)$ and $D(t, t+L)$. These two processes are independent if and only if the demand process is compound Poisson (i.e., $A(x)$ is the exponential distribution). In this special case, the marginal distribution of the inventory position can be combined with the distribution of the lead time demand in accordance with (3.20). For the joint distribution in question in the general case, we define:

$$q(t, L; x, y) dx dy \equiv P[x < I_p(t) \leq x + dx, y < D(t, t + L) \leq y + dy],$$
$$s \leq x < S, \quad y > 0, \tag{3.21}$$
$$q_0(t, L; S, 0) = P[I_p(t) = S, \ D(t, t + L) = 0], \tag{3.22}$$
$$q_1(t, L; S, y) dy = P[I_p(t) = S, \ y < D(t, t + L) \leq y + dy], \ y > 0, \tag{3.23}$$

and

$$q_2(t, L; x, 0) dx = P[x < I_p(x) \leq x + dx, \ D(t, t + L) = 0], \ s \leq x < S. \tag{3.24}$$

We now prove the following theorem.

Theorem 3.2 *The joint distribution of $I(t)$ and $D(t, t + L)$ is given* *by:*

$$q(t, L; x, y) = \sum_{n=1}^{\infty} \sum_{k=1}^{\infty} b_n(S - x) b_k(y) w(t, L; n, k)$$

$$+ \sum_{n=1}^{\infty} \sum_{k=1}^{\infty} b_n(S - x) b_k(y) \int_0^t m(t - u) w(u, L; n, k) du,$$

$$s \leq x < S, \quad y > 0, \qquad (3.25)$$

$$q_0(t, L; S, 0) = 1 - A(t + L)$$

$$+ \int_0^t m(t - u)[1 - A(u + L)] du, \qquad (3.26)$$

$$q_1(t, L; S, y) = \sum_{k=1}^{\infty} b_k(y) w(t, L; 0, k)$$

$$+ \sum_{k=1}^{\infty} b_k(y) \int_0^t m(t - u) w(u, L; 0, k) du, \quad y > 0, \, (3.27)$$

and

$$q_2(t, L; x, 0) = \sum_{n=1}^{\infty} b_n(S - x) w(t, L; n, 0)$$

$$+ \sum_{n=1}^{\infty} b_n(S - x) \int_0^t m(t - u) w(u, L; n, 0) du,$$

$$s \leq x < S, \qquad (3.28)$$

where w(u, L;n, k) is given by (2.9).

Proof For $s \leq x < S, \quad y > 0$, we obtain, again by conditioning on the renewal process of inventory cycles, that:

$$q(t, L; x, y) dx dy$$
$$= P[S - x - dx < D(t) \leq S - x, y < D(t, t + L) \leq y + dy]$$
$$+ \int_0^t m(t - u) P[S - x - dx < D(u) \leq S - x,$$
$$y < D(u, u + L) \leq y + dy] du.$$
$$(3.29)$$

On the other hand, in the same region we have:

$$P[S - x - dx < D(t) \le S - x, \; y < D(t, t + L) \le y + dy]$$

$$= \sum_{n=1}^{\infty} \sum_{k=1}^{\infty} P[S - x - dx < D(t) \le S - x, \; y < D(t, t + L) \le y + dy$$

$$|N(t) = n, \; N(t, t + L) = k]P[N(t) = n, \; N(t, t + L) = k].$$

$$(3.30)$$

The conditional probability above equals $b_n(S - x)b_k(y) \, dx \, dy$, and the probability of the condition, $P[N(t) = n, \; N(t, t + L) = k] \equiv w(t, L; n, k)$, is given by (2.9). This results in (3.25). Other results follow from similar arguments.

Time dependent and limiting distributions of inventory position, demand during a lead time and inventory on hand can be obtained from Theorem 3.2. We have already characterized the inventory position in section 3.2. For demand during a lead time, $D(t, t + L)$, we write

$$h(t, L; y)dy = P[y < D(t, t + L) \le y + dy], y > 0, \qquad (3.31)$$

and

$$h(L; y) = \lim_{t \to \infty} h(t, L; y). \qquad (3.32)$$

Corollary 3.2 *The time-dependent distribution of the lead time demand is given by:*

$$h(t, L; y) = q_1(t, L; S, y) + \sum_{n=1}^{\infty} \sum_{k=1}^{\infty} B_n(\Delta) b_k(y) w(t, L; n, k)$$

$$+ \sum_{n=1}^{\infty} \sum_{k=1}^{\infty} B_n(\Delta) b_k(y) \int_0^t m(t - u) w(u, L; n, k) du, \; y > 0,$$

$$(3.33)$$

$$P[D(t, t+L) = 0] = q_0(t, L; S, 0) + \sum_{n=1}^{\infty} B_n(\Delta) w(t, L; n, 0)$$

$$\text{(3.34)}$$

$$+ \sum_{n=1}^{\infty} B_n(\Delta) \int_0^t m(t-u) w(u, L; n, 0) du.$$

Corollary 3.3 *The limiting distribution of the lead time demand is given by:*

$$h(L, y) = \frac{1}{\mu_a} \sum_{k=1}^{\infty} b_k(y) \int_0^L [1 - A(u)][A_{k-1}(L-u) - A_k(L-u)] du, \quad y > 0,$$

$$\text{(3.35)}$$

$$P[\tilde{D}(L) = 0] = \frac{1}{\mu_a} \int_0^{\infty} [1 - A(u+L)] du, \qquad \text{(3.36)}$$

Note that in (3.33)-(3.34), the time interval of length L commences at an arbitrary epoch t, not necessarily immediately following a demand. Another immediate consequence of Theorem 3.2 is the joint distribution of inventory position and lead time demand for a stationary process. On passing to limits as $t \to \infty$ and appealing to Lemma 2.6, we find:

Corollary 3.4 *The limiting joint distribution of inventory position and demand during a lead time is given by:*

$$q(L; x, y) = \frac{r(S-x)}{\mu_a[1 + R(\Delta)]} \sum_{k=1}^{\infty} b_k(y)$$

$$\text{(3.37)}$$

$$\int_0^L [1 - A(u)][A_{k-1}(L-u) - A_k(L-u)] dy,$$

$$s \le x < S, \quad y > 0,$$

$$q_0(L; S, 0) = \frac{1}{\mu_a[1 + R(\Delta)]} \int_0^{\infty} [1 - A(u+L)] du, \qquad \text{(3.38)}$$

$$q_1(L; S, y) = \frac{r(S - x)}{\mu_a[1 + R(\Delta)]} \sum_{k=1}^{\infty} b_k(y) \int_0^L [1 - A(u)][A_{k-1}(L - u) - A_k(L - u)]du, \quad y > 0,$$

$$(3.39)$$

and

$$q_2(L; x, 0) = \frac{r(S - x)}{\mu_a[1 + R(\Delta)]} \int_0^{\infty} [1 - A(u + L)]du, \quad s \leq x < S. \quad (3.40)$$

If we compare these with the limiting marginal distributions given by (3.13)-(3.14) and (3.35)-(3.36), we discover the following.

Corollary 3.5 *The inventory position at time t, $I_p(t)$, and demand during $(t, t + L]$, $D(t, t + L)$, are asymptotically independent.*

Finally, we give the time-dependent and limiting distributions of on-hand inventory. The former is obtained from Theorem 3.2 through relationship (3.20). The latter follows either on passing to the limits or on combining the corresponding marginal distributions in accordance with $I = I_p - \tilde{D}(L)$, the stationary version of (2.20). We denote the p.d.f.'s of $I(t)$ and I by $f(t, x)$ and $f(x)$.

Corollary 3.6 *The time-dependent distribution of on-hand inventory is given by:*

$$f(t + L, x) = \begin{cases} q_1(t, L; S, S - x) + q_2(t, L; x, 0) \\ \quad + \int_x^S q(t, L; u, u - x)du, \quad s \leq x < S, \ t \geq 0, \\ \\ q_1(t, L; S, S - x) + \int_s^S q(t, L; u, u - x)du, \\ \quad\quad\quad\quad\quad\quad\quad\quad\quad x < s, \ t \geq 0, \end{cases}$$

$$(3.41)$$

$$P[I(t + L) = S] = q_0(t, L; S, 0), \quad t \geq 0. \quad (3.42)$$

Corollary 3.7 *The limiting distribution of on-hand inventory is given by:*

$$
f(x) = \begin{cases}
f_p(x)P[\tilde{D}(L) = 0] + \frac{h(L,S-x)}{1+R(\Delta)} + \displaystyle\int_x^S h(L, u - x)f_p(u)du, \\
\hspace{5cm} s \leq x < S, \\[2mm]
\frac{h(L,S-x)}{1+R(\Delta)} + \displaystyle\int_s^S h(L, u - x)f_p(u)du, \quad x < s,
\end{cases}
$$

$$ (3.43) $$

$$ P[I = S] = q_0(L; S, 0). \qquad (3.44) $$

3.3   Measures of Effectiveness

Distributions derived in the previous sections can be used to construct several measures related to the time dependent and stationary behavior of the inventory system under study. In this section we construct a number of stationary measures. We also derive the stationary distribution of the customer waiting time by independent arguments. (Recall that s is allowed to be negative.)

3.3.1   Means and Variances

Starting with the inventory position, from (3.13)-(3.14) we have:

$$ E[I_p] = s + \frac{\Delta + \bar{R}(\Delta)}{1 + R(\Delta)}, \qquad (3.45) $$

and

$$ Var[I_p] = \frac{\Delta^2 + 2\bar{\bar{R}}(\Delta)}{1 + R(\Delta)} - [\frac{\Delta + \bar{R}(\Delta)}{1 + R(\Delta)}]^2, \qquad (3.46) $$

where $\bar{R}(x) = \int_0^x R(u)du$ and $\bar{\bar{R}}(x) = \int_0^x \bar{R}(u)du$.

Similarly, for demand during a lead time we find from (3.35)-(3.36):

$$E[\tilde{D}(L)] = \frac{\mu_b}{\mu_a} L, \tag{3.47}$$

$$Var[\tilde{D}(L)] = \frac{\mu_b^{(2)}}{\mu_a}[L + 2\bar{R}(L)] - (\frac{\mu_b L}{\mu_a})^2. \tag{3.48}$$

On account of $I = I_p - \tilde{D}(L)$, we then have

$$E[I] = E[I_p] - E[\tilde{D}(L)]. \tag{3.49}$$

And, since I_p and $\tilde{D}(L)$ are independent (cf. Corollary 3.5)

$$Var[I] = Var[I_p] + Var[\tilde{D}(L)]. \tag{3.50}$$

In addition, the expected on hand positive stock is:

$$E[I^+] \equiv \int_0^S x \, dF(x) = \frac{\bar{H}(L,S) + \int_{[s]^+}^S r(S-u)\bar{H}(L,u)du}{1 + R(\Delta)}, \tag{3.51}$$

where $[s]^+ = max(s, 0)$, and the expected outstanding backorders is given by:

$$E[I^-] \equiv \int_{-\infty}^0 x \, dF(x) = E[I] - E[I^+]. \tag{3.52}$$

These follow from (3.43)-(3.44).

3.3.2   The Fill Rate

The stationary probability that on hand inventory is positive is obtained from (3.43)-(3.44) to be:

$$P[I > 0] \equiv 1 - F(0) = \frac{H(L,S) + \int_{[s]^+}^S r(S-u)H(L,u)du}{1 + R(\Delta)}. \tag{3.53}$$

This measure is generally referred to as the *fill rate*. Clearly, $F(0)$ is the stockout probability.

Behavior of $1 - F(0)$ as Δ varies will be of some interest in the sequel both for an arbitrary s and for $s = 0$. Especially in the latter case, it is intuitively clear that a larger Δ would increase the lengths of time during which $I > 0$. It is also clear in this case, provided that there is no more than one outstanding order during a lead time, that the lengths of time during which $I < 0$ are all equal to L independently of Δ. Consequently, $1 - F(0)$ should be increasing in Δ. Unfortunately, when there are more than one outstanding order during a lead time, stockout spells may also increase in length as Δ is increased, depending on the characteristics of the demand process. Thus, $1 - F(0)$ may decrease in Δ over one or more intervals. The situation is even more complicated when $s \neq 0$. It turns out for $L = 0$, $s < 0$, for example, that $1 - F(0)$ is monotone increasing in Δ only under a rather restrictive condition on the demand process. At this point, we note the following result.

Lemma 3.1 $1 - F(0)$ *is monotone increasing in* Δ *if* $1 + R(\Delta)$ *is log-concave (i.e.,* $log(1 + R(\Delta))$ *is concave).*

Proof We have $I = I_p - \tilde{D}(L)$ where, according to Corollary 3.5, I_p and $\tilde{D}(L)$ are independent. Clearly, $P[I > 0] = P[I_p > \tilde{D}(L)]$. On the other hand, it follows from (3.13)-(3.14) that

$$P[I_p > x] = \frac{1 + R(S - x)}{1 + R(\Delta)}, \quad x \geq s. \qquad (3.54)$$

It is seen that this measure is increasing in Δ if $r(S - x)/[1 + R(S - x)] > r(\Delta)/[1 - R(\Delta)]$ or $d/d\Delta \; log[1 + R(S - x)] > d/d\Delta \; log[1 + R(\Delta)]$ which holds if $1 + R(x)$ log-concave.

Implications of the log-concavity of $1 + R(x)$ to optimization problems in inventory systems and related areas were examined by Stidham (1975, 1986). (Recall that $\mu_a[1 + R(\Delta)]$ is the expected cycle length.) Although log-concavity is a weaker assumption than concavity, there is no distribution of practical interest with log-concave but not concave renewal function.

3.3.3 Customer Waiting Time

An important service level measure for systems with backlogging is the customer waiting time. Here we consider the actual waiting time of a customer who arrives at a given time. We define this time, in the presence of continuously distributed batch demands, as measured from the time of arrival of a customer to the time at which his demand is completely filled. Working with a discrete batch size distribution, Kruse (1981) uses a more refined waiting time concept, defined separately for each unit of the batch demanded. In what follows, we shall allow s to be negative; Kruse (1981) requires in a discrete state process that $s \geq -1$ which in turn implies that no one waits more than L.

Let $W(t)$ denote the waiting time of a customer who arrives at time t and write $\lim_{t \to \infty} P[W(t) \leq y] \equiv P[W \leq y]$. We now prove the following results:

Theorem 3.3

$$P[W \leq y] = \tfrac{1}{1+R(\Delta)} \sum_{n=0}^{\infty} [A_n(L-y) - A_{n+1}(L-y)]$$

(3.55)

$$P[W = L] = \frac{1}{1 + R(\Delta)} \begin{array}{l} [B_{n+1}(S) + \int_{[s]+}^{S} r(S-u)B_{n+1}(u)du], \quad y < L, \\[2mm] [[1-B(S)] + \int_{[s]+}^{S} r(S-u)[1-B(u)]du], \end{array}$$

(3.56)

and

$$P[L < W \leq y] = \begin{cases} 0, \quad s \geq 0 \\[2mm] \dfrac{1}{1+R(\Delta)}\{R(\Delta) - R(S) \\[2mm] \quad - \sum_{n=0}^{\infty} [A_n(y-L) - A_{n+1}(y-L)] \\[2mm] \quad \int_{s}^{0} r(S-u)B_n(u-s)du\}, \quad y > L, \quad s \leq 0. \end{cases}$$

(3.57)

Proof First, observe that a customer who arrives at time t, demanding X, and who finds $I_p(t-) \geq X$, will wait $\leq y < L$ if and only if $I_p(t + y - L)$, which will be available for issue by time $t + y$, less $D(t + y - L, t-)$ is larger than X. In symbols:

$$\{W(t) \leq y\} \Leftrightarrow \{I_p(t+y-L) - D(t+y-L, t-) \geq X\}, \quad y < L. \quad (3.58)$$

Note that since t is a demand point, looking backward from t, $D(t + y - L, t-)$ has the same distribution as $D(L - y)$ which in turn is given by (3.4). It follows, for a stationary process, that:

$$P[W \leq y | I_p = u, X = x] = P[D(L - y) \leq u - x], \quad (3.59)$$

where $y < L$, $[s]^+ \leq u \leq S$ and $x \leq u$. Therefore:

$$P[W \leq y | I_p(u)] = \int_0^u P[D(L - y) \leq u - x] b(x) dx, \quad (3.60)$$

and

$$P[W \leq y] = \frac{1}{1 + R(\Delta)} \{ \int_0^S P[D(L - y) \leq S - x] b(x) dx$$

$$+ \int_{u=[s]^+}^S r(S - u) \int_{x=0}^u P[D(L - y) \leq u - x] b(x) dx \, du \}, \quad y < L,$$

$$(3.61)$$

where we used (3.13)-(3.14). The result (3.55) follows on substituting (3.4).

To complete the construction of the distribution, we now distinguish two cases. If $s \geq 0$ so that no customer waits more than L, then $P[W \leq L] = 1$. This observation or the following argument establishes the probability concentration at L.

$$P[W = L | I_p = u] = 1 - B(u), \quad u \geq 0, \quad (3.62)$$

$$P[W = L] = \frac{1}{1 + R(\Delta)} \{ [1 - B(S)] + \int_s^S r(S - u)[1 - B(u)] du \},$$

$$s \geq 0.$$

$$(3.63)$$

For $s < 0$, a similar argument produces:

$$P[W = L] = \frac{1}{1 + R(\Delta)}\{[1 - B(\Delta)] + \int_s^S r(S - u)[1 - B(u - s)]du\}$$

$$= \frac{1}{1 + R(\Delta)}\{[1 - B(\Delta)] + \int_0^\Delta r(\Delta - u)[1 - B(u)]du\}$$

$$= \frac{1}{1 + R(\Delta)}, \quad s \leq 0. \tag{3.64}$$

Note that (3.56) includes both (3.63) and (3.64).

For $s < 0$, we also have the possibility that $W > L$. This occurs when an arrival at time t finds $0 > I_p(t+) > s$. Such a customer would have to wait until the next order placement time plus L. Accordingly:

$$P[L < W \leq y] = \tfrac{1}{1+R(\Delta)}\int_s^0 r(S - u)P[D(y - L) > u - s]du,$$

$$y > L, \quad s \leq 0. \tag{3.65}$$

(3.57) follows with (3.4). This completes the proof.

In closing this section, we note that for $s < 0$, (3.55)-(3.56) imply that

$$P[W \leq L] = \frac{1 + R(S)}{1 + R(\Delta)}, \quad s \leq 0, \tag{3.66}$$

and, according to (3.57):

$$P[L < W < \infty] = \frac{R(\Delta) - R(S)}{1 + R(\Delta)}, \quad s \leq 0. \tag{3.67}$$

As we noted before, $P[W \leq L] = 1$ for $s \geq 0$.

3.3.4   The Cost Rate Function

The most common measure of effectiveness used in relation to the model under consideration is the stationary cost rate function that is obtained by imposing a cost structure on the system in steady state.

We choose a simple representation involving a fixed ordering cost of K dollars per order, and linear holding and shortage costs with rates h and p dollars per unit per unit time, respectively. The expected total cost per unit time in steady state can be expressed as:

$$E(s, \Delta) = \frac{K}{\mu_a[1 + R(\Delta)]} + h \int_0^S x \, dF(x) - p \int_{-\infty}^0 x \, dF(x)$$

$$= \frac{k}{1 + R(\Delta)} + (h + p)E[I^+] - pE[I], \qquad (3.68)$$

where $k \equiv K/\mu_a$, and $E[I^+]$ and $E[I]$ are given by (3.51) and (3.49), respectively. For future reference, we rewrite $E(s, \Delta)$ in the following more convenient form:

$$E(s, \Delta) = \frac{N(s, \Delta)}{1 + R(\Delta)} + \frac{\mu_b}{\mu_a} pL, \qquad (3.69)$$

where

$$N(s, \Delta) = \; k - p[S + sR(\Delta) + \bar{R}(\Delta)] + (p + h)[\bar{H}(L, S) \\ + \int_{[s]^+}^S r(S - u)\bar{H}(L, u)du]. \qquad (3.70)$$

3.4 Special Cases

In this section we discuss two important special cases of the model presented above: systems with unit demands (i.e., each demand is for a single item) and systems under periodic review.

3.4.1 Unit Demands

The first systematic investigation of an inventory system under unit demands separated by i.i.d. interdemand times was given by Finch (1961) as a special case of a model with random lead times, and by Beckman (1961) as a special case of a model with discrete random batch sizes. An independent study of the model under the assumption of immediate deliveries is found in Sivazlian (1974). The model is also included, at least formally, as a special case of the more general systems studied in the previously cited references.

Consistent with the earlier representations, we will use the convention that whenever $I_p(t) = s - 1$, a replenishment order of size $\Delta+1$ is placed, thus raising the inventory position to S where s, S, and $\Delta = S - s$ are now integers; s is allowed to be negative as before. The alternative would be to place orders of size Δ whenever $I_p(t) = s$. However, this is inconsistent with our earlier convention of ordering $S - I_p(t)$ whenever $I_p(t) < s$ in the case of a continuous-state process. In either case, note that the *reorder point, order-up-to-level* policy is reduced, effectively, to a *reorder-point, order-quantity* policy.

Another feature that is also unique to the case of unit demands is that the stationary distribution of inventory position turns out to be uniform. Together with a constant procurement lead time, this implies, as we show in the sequel, a unimodal cost rate function that facilitates the determination of optimal policies. These characteristics, and the prevalence of the model in practice, such as in the retail business of consumer durables, justify a separate treatment. In this part, we provide, without proof, the unit-demand-size versions of the results obtained so far. Optimality conditions, optimal policies, and approximations will be discussed in the sequel, again as special cases of the corresponding more general findings.

We first note that the representation of unit batch sizes by $b(1) = 1$ immediately implies that $r(j) = 1$, $j = 1, 2, ...$, and $R(j) = j$ for the renewal density and function defined through batch sizes.

For the time dependent and stationary joint distributions of inventory position and demand during a lead time, let $q(t, L; n, m) = P[I_p(t) = n, D(t, t+L) = m]$, and $q(L; n, m) = \lim_{t \to \infty} q(t, L; n, m) \equiv P[I_p = n, \bar{D}(L) = m]$. We have:

Theorem 3.4

$$q(t, L; n, m) = w(t, L; S - n, m) + \int_0^t m(t - u)w(u, L; S - n, m)du,$$
$$(3.71)$$

where $w(u, L; n, m)$ is given by (2.16) and $m(t)$ is the solution of

$$m(t) = a_{\Delta+1}(t) + \int_0^t a_{\Delta+1}(t - u)m(u)du. \qquad (3.72)$$

Also,

$$q(n, m) = \begin{cases} \dfrac{1}{\mu_a[1+\Delta]} \displaystyle\int_0^L [1 - A(u)][A_{m-1}(L - u) - A_m(L - u)]du, \\ \qquad\qquad\qquad\qquad\qquad\qquad m \geq 1 \\ \\ \dfrac{1}{\mu_a[1+\Delta]} \displaystyle\int_0^\infty [1 - A(u + L)]du, \quad m = 0. \end{cases}$$
$$(3.73)$$

Proof of (3.71) follows the same arguments as in Theorem 3.2. Recall that $w(u, L; n, m) = P[N(u) = n, N(u, u + L) = m]$ where $N(u, u + t) \equiv D(u, u+t)$ is the total demand during $(u, u+t]$. As before, if u is an arrival point, we write $D(u, u+t) \equiv D(t)$; this has the distribution: $P[D(t) = n] = A_n(t) - A_{n+1}(t)$, $n = 0, 1, \ldots$. We also note that $m(t)$ is the renewal density of the renewal process of inventory cycles; $a_{\Delta+1}(t)$, the $\Delta + 1$st convolution of the interarrival density function, is the interval density of this renewal process. Correspondingly, the expected cycle length is $\mu_a[1 + \Delta]$.

As corollaries to Theorem 3.4, we next give the distributions of inventory position and demand during a lead time.

Corollary 3.8 *Let $P[I_p(t) = n] \equiv f_p(t, n)$, then*

$$f_p(t, n) = \begin{cases} P[D(t) = S - n] + \displaystyle\int_0^t m(t - u) \\ \qquad\qquad P[D(u) = S - n]du, \quad s \leq n < S, \\ \\ 1 - A(t) + \displaystyle\int_0^t m(t - u)[1 - A(u)]du, \quad n = S. \end{cases}$$
$$(3.74)$$

and

$$f_p(n) = \frac{1}{\Delta + 1}, \quad s \le n \le S. \tag{3.75}$$

Thus the stationary distribution of inventory position is uniform over the set $\{s, s+1, ..., S\}$. If the control policy called for ordering whenever $I_p(t) = s$, rather than when $I_p(t) = s - 1$ as above, this distribution would have been uniform over the set $\{s+1, s+2, ..., S\}$. Richards (1975) proved that if demands are generated by a renewal process, then $f_p(n)$ is uniform under an (s, S) policy if and only if demands are of unit size.

Corollary 3.9 *Time dependent and stationary distributions of demand during a lead time are given, respectively, by:*

$$h(t, L; m) = \sum_{n=0}^{\Delta} [w(t, L; n, m) + \int_0^t m(t - u)w(u, L; n, m)du],$$
$$t \ge 0, \quad m = 0, 1, ..., \tag{3.76}$$

and

$$h(L, m) = \begin{cases} \frac{1}{\mu_a} \int_0^L [1 - A(u)][A_{m-1}(L - u) - A_m(L - u)]du, & m \ge 1 \\[2ex] \frac{1}{\mu_a} \int_0^\infty [1 - A(u + L)]du, & m = 0. \end{cases} \tag{3.77}$$

For later reference, we also note the c.d.f. corresponding to the limiting density (3.77) to be:

$$H(L, m) = 1 - \frac{1}{\mu_a} \int_0^L [1 - A(u)]A_m(L - u)du, \quad m \ge 0. \tag{3.78}$$

In all these expressions the time interval of length L commences at an arbitrary epoch t, not necessarily at a demand point. We also see, from (3.73), (3.75), and (3.77) that $I_p(t)$ and $D(t, t + L)$ are asymptotically independent.

The time dependent distribution of on-hand inventory is obtained from Theorem 3.4 and the relation $I(t + L) = I_p(t) - D(t, t + L)$. The limiting distribution follows either on passing to the limits or on combining the corresponding marginal distributions in accordance with $I = I_p - \tilde{D}(L)$. We have:

Corollary 3.10

$$f(t+L, n) = \sum_{k=max(n,s)}^{S} q(t, L; k, k-n), \quad t \geq 0, \quad -\infty < n \leq S, \quad (3.79)$$

and

$$f(n) = \frac{1}{\Delta + 1} \sum_{k=max(n,s)}^{S} h(L, k - n), \quad -\infty < n \leq S. \quad (3.80)$$

In terms of measures of effectiveness, most of the results reported in part 3.3 can be easily specialized to the case of unit demands on substituting $R(j) = j$. For the fill rate, we have:

$$1 - F(0) = \frac{1}{\Delta + 1} \sum_{k=[s-1]^+}^{S-1} H(j). \quad (3.81)$$

And for the expected inventory levels $E[I]$ and $E[I^+]$ we obtain

$$E[I] = s + \frac{\Delta}{2} - \frac{L}{\mu_a}, \quad (3.82)$$

and

$$E[I^+] = \frac{1}{1 + \Delta} [\sum_{n=0}^{S} nH(S - n) - \sum_{n=0}^{[s-1]^+} nH(s - n - 1)]. \quad (3.83)$$

It follows, for the cost rate function, that:

$$E(s, \Delta) = \frac{N(s, \Delta)}{1 + \Delta} + \frac{pL}{\mu_a} \qquad (3.84)$$

where

$$N(s, \Delta) = k - p(s + \tfrac{\Delta}{2})(\Delta + 1)$$
$$+ (p + h)[\sum_{n=0}^{S} nH(S - n) - \sum_{n=0}^{[s-1]^+} nH(s - n - 1)]. \qquad (3.85)$$

In the next chapter we prove that $E(s, \Delta)$ is unimodal.

Finally, the stationary distribution of the waiting time of an arrival turns out to be:

$$P[W \le y] = \frac{1}{1 + \Delta} \sum_{n=[s]^+}^{S} [1 - A_n(L - y)], \quad y \le L, \qquad (3.86)$$

and

$$P[L < W \le y] = \tfrac{1}{1+\Delta}\{-s - \sum_{n=0}^{-s-1}(n + s)$$
$$[A_n(y - L) - A_{n+1}(y - L)]\}, \quad s < 0, \quad y > L. \qquad (3.87)$$

3.4.2   The Periodic Review Model

Recall from Chapter 1 that in the periodic review version of the model the lead time is an integral multiple of the period length which in turn is regarded as unity. Demands in successive periods are i.i.d. random variables with c.d.f. $B(x)$ which is assumed to be continuous with p.d.f. $b(x)$. If $D(L)$ is the lead time demand, we have:

$$P[D(L) \le x] = B_L(x), \quad x \ge 0, \quad L \ge 0. \qquad (3.88)$$

with the corresponding density function $b_L(x)$. The lead time demand is always characterized by (3.88) as every lead time starts and ends at the beginning (or end) of a period.

According to the observations made in Chapter 1, regarding the equivalency of the periodic review model to a properly specialized continuous review model, if, in the foregoing we take $H(L, x) \equiv B_L(x)$, $h(L, x) = b_L(x)$, $A(x) = 1$ if $x \geq 1$, $A(x) = 0$ if $x < 1$, and replace L by $L + 1$, we should obtain the corresponding results for the periodic review system. Although the specialization to constant interdemand times violates the continuity assumption on $A(x)$, this flexibility can be added to the formalism.

Using the transformations noted above, the stationary distribution of on hand inventory can be obtained from (3.43). Clearly, $P[I = S] = P[D(L) = 0] = 0$, and:

$$f(x) = \tfrac{1}{1+R(\Delta)}[b_{L+1}(S - x) + \int_{max(x,s)}^{S} r(S - u)b_{L+1}(u - x)du],$$

$$-\infty < x \leq s.$$
$$(3.89)$$

The renewal function $R(x)$ and its density $r(x)$ now correspond to the renewal process formed by the periodic demand sequence with common p.d.f. $b(x)$. Putting $L = 0$ in the above, the p.d.f. of the inventory position turns out to be:

$$f_p(x) = \begin{cases} \dfrac{r(S - x)}{1 + R(\Delta)}, & s \leq x \leq S, \\[3mm] \dfrac{v_\Delta(s - x)}{1 + R(\Delta)}, & x \leq s, \end{cases} \qquad (3.90)$$

where $v_\Delta(u)$ is the p.d.f. of the *forward recurrence time* given by (2.19).

For measures of effectiveness, the results of section 3.3 remain valid under the above transformations, and with $\mu_a \equiv 1$, $E[\tilde{D}(L)] \equiv E[D(L)] \equiv \mu_b L$, $Var[\tilde{D}(L)] \equiv Var[D(L)] = \sigma_b^2 L$, $H(L, x) \equiv B_{L+1}(x)$ and $\bar{H}(L, x) \equiv \bar{B}_{L+1}(x) = \int_0^x B_{L+1}(u)du$. In particular, we rewrite below the expression for the cost rate function and provide an equivalent form for it for later reference. We have, from (3.69)-(3.70) that

$$E(s, \Delta) = \frac{N(s, \Delta)}{1 + R(\Delta)} + \mu_b p(L + 1), \qquad (3.91)$$

where

$$N(s, \Delta) = K - p[S + sR(\Delta) + \bar{R}(\Delta)]$$
$$+ (p + h)[\bar{B}_{L+1}(S) + \int_{[s]+}^{S} r(S - u)\bar{B}_{L+1}(u)du]. \qquad (3.92)$$

An alternative form of $N(s, \Delta)$ is obtained by noting that the expression inside the last brackets can be written as:

$$\begin{cases} \int_0^S r(S - u)\bar{B}_L(u)du, & s \leq 0, \\[3mm] \int_0^s v_\Delta(s - x)\bar{B}_L(u)du + \int_s^S r(S - u)\bar{B}_L(u)du, & s \geq 0. \end{cases} \qquad (3.93)$$

Chapter 4

Optimality Conditions

This chapter is mainly concerned with the optimality conditions relative to the cost rate function $E(s, \Delta)$. We also establish some useful bounds on the optimal policy. Most of the results are again given for the general model under continuous review which includes the periodic review model as a special case. Results for two other special cases, involving immediate deliveries and unit demands, are provided separately in section 4.2.

A sufficient condition for the unimodality, in general, of $E(s, \Delta)$, is that $1 + R(x)$ be log-concave (Stidham, 1977, 1986). This excludes large classes of demand distributions that arise in practice. Also it is not a necessary condition for $E(s, \Delta)$ to be unimodal except when $L = 0$. The chapter provides a simplified exposition of these facts. Log-concavity of $1 + R(x)$ may appear to be an unexpectedly strong condition for $E(s, \Delta)$ to be unimodal. But, in fact, when this condition holds, the cost rate function is unimodal for any set of values for cost parameters K, h and p. Otherwise, it is easy to establish, for example, that $E(s, \Delta)$ is unimodal if K is large enough.

One suspects that the conditions that induce $E(s, \Delta)$ to be unimodal in the above sense should also result in other nice properties.

We prove in the sequel that the unimodality in question is equivalent to the property that the fill rate $1 - F(0)$ is monotone increasing in Δ, which in turn is equivalent to the property that the optimal s for a given Δ is monotone decreasing in Δ. Any one of these properties may appear to be self-evident without any condition; unfortunately, such is not the case.

4.1   The General Case

First, we note that for any given $\Delta \geq 0$, $E(s, \Delta)$ is convex in s, for we have from (3.69) that:

$$\frac{\partial^2 E(s, \Delta)}{\partial s^2} = \frac{p+h}{1+R(\Delta)}[h(L, s) + \int_{[s]+}^{S} r(S - u)h(L, u)du] > 0. \quad (4.1)$$

However, $E(s, \Delta)$ is not, in general, convex or pseudoconvex on $\Delta \geq 0$. The necessary conditions for optimality, $\partial E/\partial s = 0$, $\partial E/\partial \Delta = 0$, can be written as:

$$\frac{1}{1+R(\Delta)}[H(L, S) + \int_{[s]+}^{S} r(S - u)H(L, u)du] = \frac{p}{p+h}, \quad (4.2)$$

and

$$\frac{p[\Delta + \bar{R}(\Delta)]}{p+h} - \int_{[s]+}^{S} [1 + R(S - u)]H(L, u)du = \frac{k}{p+h}. \quad (4.3)$$

For $\Delta \geq 0$ given, let $s^o(\Delta)$ minimize $E(s, \Delta)$; that is:

$$E(s^o(\Delta), \Delta) \leq E(s, \Delta), \quad -\infty < s < \infty. \quad (4.4)$$

By convexity in s of $E(s, \Delta)$, $s^o(\Delta)$ is unique and is determined by (4.2). Let

$$I(\Delta) = \frac{1}{1+R(\Delta)}[H(L, \Delta) + \int_0^{\Delta} r(\Delta - u)H(L, u)du], \quad (4.5)$$

and define Δ_1 by:

$$I(\Delta_1) = \frac{p}{p+h}. \quad (4.6)$$

We have the following result for the sign of $s^o(\Delta)$.

Lemma 4.1 *a)* $s^o(\Delta) > (=)0$ *if and only if* $\Delta < (=)\Delta_1$.

b) Δ_1 *is unique if and only if*

$$\frac{r(x)}{1 + R(x)} \geq \frac{\displaystyle\sum_{n=1}^{\infty}(1 - c_n)b_n(x)}{1 + \displaystyle\sum_{n=1}^{\infty}(1 - c_n)B_n(x)}, \tag{4.7}$$

where

$$c_n = \frac{1}{\mu_a}\int_0^L [1 - A(u)][1 - A_n(L - u)]du, \quad n \geq 1. \tag{4.8}$$

Proof The Lemma is an extension of Lemma 1 of Şahin (1982). It can be seen that $E(s, \Delta)$ is increasing in s for s large enough. It follows, due to convexity of $E(s, \Delta)$ in s that if $\partial/\partial s\, E(s, \Delta)|_{s=0} > (=)(<)0$, that is if $I(\Delta) > (=)(<)p/(p + h)$, then $s^o(\Delta) < (=)(>)0$ and conversely. Clearly, Δ is unique for every $p/(p + h) \geq 0$ if and only if $I(\Delta)$ is monotone increasing on $\Delta \geq 0$. The inequality (4.7) is the necessary and sufficient condition for $I(x)$ to be monotone increasing in x. This is obtained by using $(3.35) - (3.36)$ in $d/dx\, I(x) \geq 0$. The latter can be written as:

$$\frac{r(x)}{1 + R(x)} \leq \frac{h(x) + \displaystyle\int_0^x r(x - u)h(u)du}{H(x) - H(0) + \displaystyle\int_0^x R(x - u)h(u)du}$$

$$= \frac{\displaystyle\sum_{n=1}^{\infty}(c_n - c_{n-1})[b_n(x) + \int_0^x r(x - u)b_n(u)du]}{\displaystyle\sum_{n=1}^{\infty}(c_n - c_{n-1})[B_n(x) + \int_0^x R(x - u)b_n(u)du]}$$

$$= \frac{\sum_{n=1}^{\infty} (c_n - c_{n-1}) \int_0^x r(x-u) b_{n-1}(u) du}{\sum_{n=1}^{\infty} (c_n - c_{n-1}) \int_0^x R(x-u) b_{n-1}(u) du}$$

$$= \frac{\sum_{n=1}^{\infty} (c_n - c_{n-1}) \sum_{j=1}^{\infty} b_{j+n-1}(x)}{\sum_{n=1}^{\infty} (c_n - c_{n-1}) \sum_{j=1}^{\infty} B_{j+n-1}(x)}$$

$$= \frac{\sum_{n=1}^{\infty} c_n b_n(x)}{\sum_{n=1}^{\infty} c_n B_n(x)} \qquad (c_0 = 0)$$

$$= \frac{r(x) - \sum_{n=1}^{\infty} (1 - c_n) b_n(x)}{R(x) - \sum_{n=1}^{\infty} (1 - c_n) B_n(x)}.$$

The inequality (4.7) follows.

Condition (4.7) is clearly satisfied for the special case of $L = 0$ which results in $c_n = 0$ and (4.7) holds as equality. Note also that $I(\Delta) \equiv \bar{F}(0)$ for an inventory system operating under a $(0, \Delta)$ policy where $\bar{F}(0)$ is the *fill rate* given by (3.53). According to Lemma 3.1, $\bar{F}(0)$ is increasing in Δ (for any s) if $1 + R(x)$ is log-concave on $x \geq 0$. Therefore, $I(\Delta)$ is increasing in Δ and (4.7) is satisfied under the same condition. Finally, for the periodic review version of the model, $A(x) = 1$, $x \geq 1$, $A(x) = 0$, $x < 1$, and $L \to L + 1$ imply $c_n = 0$, $n \leq L$, $c_n = 1$, $n > L$. In this case (4.7) holds when $L = 0$. In general (4.7) poses an interesting characterization problem that has not yet been completely resolved.

Next, we record another useful result that was also noted in Şahin (1982).

Lemma 4.2 *For $I(\Delta) \leq p/(p+h)$ we have:*

$$H(L, s^o(\Delta) + \Delta) \geq p/(p+h) \geq H(L, s^o(\Delta)). \qquad (4.9)$$

Proof For $I(\Delta) \leq p/(p+h)$, we have $s^o(\Delta) \geq 0$ by Lemma 4.1. On comparing (3.53) with (4.3), we find that under any policy of the form $(s^o(\Delta), \Delta + s^o(\Delta))$, $\Delta \geq 0$, we have

$$F(0) \equiv P[I \leq 0] = h/(p+h). \qquad (4.10)$$

On the other hand, we can state, for any $s \geq 0$, $\Delta \geq 0$ that

$$\{\tilde{D}(L) \leq s\} \Rightarrow \{I > 0\} \Rightarrow \{\tilde{D}(L) \leq s + \Delta\}. \qquad (4.11)$$

Both implications follow from $I = I_p - \tilde{D}(L)$ and $s \leq I_p \leq S$. The lemma follows from (4.10) and (4.11).

We are now in a position to establish Δ_1, under the condition of its uniqueness, as a bound on Δ^* and S^*. For this purpose we define:

$$M(\Delta) = k + \int_0^\Delta [1 + R(\Delta - u)]\{hH(L, u) - p[1 - H(L, u)]\} du. \qquad (4.12)$$

Theorem 4.1 *Assume that condition (4.7) holds, then $\Delta^* > (=)\Delta_1$, and $S^* > (=)\Delta_1$ if and only if $M(\Delta_1) > (=)0$.*

Proof We first show that $M(\Delta_1) \geq 0$ implies $\Delta^* \geq \Delta_1$. This part of the proof is based on the observation that $N(s^o(\Delta), \Delta)$ in (3.70) is monotone decreasing on $0 \leq \Delta \leq \Delta_1$. By Lemma 4.1, $s^o(\Delta) \geq 0$ in this region and is determined by (4.2). For simplicity, we write $s^o(\Delta) \equiv s$ and $ds^o(\Delta)/d\Delta \equiv s'$. We have:

$$\frac{d}{d\Delta} N(s, \Delta) = -p\{(1 + s')[1 + R(\Delta)] + sr(\Delta)\}$$

$$+ \quad (p+h)\{r(\Delta)\bar{H}(l,s) + (1+s')$$

$$[H(L,S) + \int_s^S r(S-u)H(L,u)du]\}$$

$$= \quad r(\Delta)[(p+h)\bar{H}(L,S) - ps]$$

$$\leq \quad sr(\Delta)[(p+h)H(L,s) - p] \leq 0, \qquad (4.13)$$

where we used (4.2), the fact that $\bar{H}(L,s) \leq sH(L,s)$, and Lemma 4.2.

Since $R(\Delta)$ is monotone increasing on $\Delta \geq 0$, and, as shown above, $N(s,\Delta)$ is monotone decreasing on $0 \leq \Delta \leq \Delta_1$, $E(s,\Delta)$ would be monotone decreasing on $0 \leq \Delta \leq \Delta_1$ if $N(s,\Delta) \geq 0$ in this region. As $s^o(\Delta_1) = 0$ by Lemma 4.2, we have $N(s^o(\Delta_1),\Delta_1) = M(\Delta_1)$ where $M(\Delta)$ is given by (4.12). Since $M(\Delta_1) \geq 0$ by assumption, it follows that $\Delta^* \geq \Delta_1$ and, by Lemma 4.1, $s^* \leq 0$.

Next we show that $\Delta^* \geq \Delta_1$ implies $M(\Delta_1) \geq 0$. For $\Delta^* \geq \Delta_1$, we have $s^* \leq 0$ by Lemma 4.1. If we use (4.3) in (3.69) we find:

$$E(s^*,\Delta^*) = p(\frac{\mu_b}{\mu_a}L - s^*) \geq \frac{\mu_b}{\mu_a}pL. \qquad (4.14)$$

On the other hand, consider

$$E(0,\Delta_1) = \frac{M(\Delta_1)}{1 + R(\Delta_1)} + \frac{\mu_b}{\mu_a}pL, \qquad (4.15)$$

$M(\Delta_1) < 0$ implies $E(0,\Delta_1) < E(s^*,\Delta^*)$ which is a contradiction; we must have $M(\Delta_1) \geq 0$. We note that if $M(\Delta_1) = 0$, then $s^* = 0$ and $\Delta^* = \Delta_1$ are optimal by (4.14)-(4.15). Conversely, if $(0,\Delta_1)$ is the optimal policy, then (4.3) in (3.70) implies $M(\Delta_1) = 0$.

We have thus far proved that $\Delta^* > (=)\Delta_1$ if and only if $M(\Delta_1) > (=)0$. To complete the proof of the theorem, we now show that $\Delta^* >$

$(=)\Delta_1$ if and only if $S^* > (=)\Delta_1$. For this purpose, assume first that $\Delta^* \geq \Delta_1$. We have $s^* \leq 0$ by Lemma 4.1. From (4.2), Δ^* satisfies

$$\frac{Q(S^*)}{1 + R(\Delta^*)} = \frac{p}{p + h}, \qquad (4.16)$$

and, from Lemma 4.2, Δ_1 is determined by

$$\frac{Q(\Delta_1)}{1 + R(\Delta_1)} = \frac{p}{p + h}, \qquad (4.17)$$

where

$$Q(x) = H(L, x) + \int_0^x r(x - u)H(L, u)du. \qquad (4.18)$$

It is seen that $Q(x)$ is monotone increasing in x, for

$$\frac{d}{dx}Q(x) = h(L, x) + r(x)H(L, 0) + \int_0^x r(x - u)h(L, u)du \geq 0. \ (4.19)$$

Consequently, by (4.16)-(4.17), $\Delta^* > (=)\Delta_1$ implies $S^* > (=)\Delta_1$.

Finally, assume $\Delta^* < \Delta_1$. We have $s^* > 0$ by Lemma 4.1. For $\Delta < \Delta_1$, let $s^o(\Delta) > 0$ be determined by (4.2). Using implicit differentiation, we find:

$$\frac{d}{d\Delta}s^o(\Delta) = -\frac{q(s, \Delta) + r(\Delta)[H(L, s) - p/(p + h)]}{q(s, \Delta)} \qquad (4.20)$$

where

$$q(s, \Delta) = h(L, S) + \int_s^S r(S - u)h(L, u)du. \qquad (4.21)$$

Since $S = \Delta + s$, letting $S^o(\Delta) = \Delta + s^o(\Delta)$, we have:

$$\frac{d}{d\Delta}S^o(\Delta) = 1 + \frac{d}{d\Delta}s^o(\Delta)$$

$$= \frac{r(\Delta)[p/(p + h) - H(L, s)]}{q(s, \Delta)} \geq 0 \qquad (4.22)$$

where the inequality follows from Lemma 4.2. Consequently $S^o(\Delta)$ is increasing in Δ on $0 \leq \Delta \leq \Delta_1$. On the other hand, we have by

Lemma 4.1 that $s^o(\Delta_1) = 0$, therefore $S^o(\Delta_1) = \Delta_1$. It follows that $S^o(\Delta) < \Delta_1$ on $0 \leq \Delta < \Delta_1$ and $S^* < \Delta_1$. This completes the proof of the theorem.

Theorem 4.1 and Lemma 4.1 together imply that if $M(\Delta_1) > 0$ then $s^* < 0$ and $\Delta^* > S^* > \Delta_1$, if $M(\Delta_1) < 0$ then $s^* > 0$ and $\Delta^* < S^* < \Delta_1$, and if $M(\Delta_1) = 0$ then $s^* = 0$ and $\Delta^* = \Delta_1$. Thus Δ_1 is either a lower or an upper bound on Δ^* and S^*. These observations have important implications to the computation and approximation of optimal policies. We now give a sufficient optimality condition that is also useful in computations when $E(s, \Delta)$ is not unimodal.

Theorem 4.2 *Assume condition 4.7 holds. If $(\hat{s}, \hat{\Delta})$ and $(\tilde{s}, \tilde{\Delta})$, $\hat{s} \geq \tilde{s}$, both satisfy the necessary optimality conditions, then $E(\hat{s}, \hat{\Delta}) \leq E(\tilde{s}, \tilde{\Delta})$.*

Proof First, assume $M(\Delta_1) \geq 0$ so that $s^* \leq 0$. Let $(\hat{s}, \hat{\Delta})$ and $(\tilde{s}, \tilde{\Delta})$, $\tilde{s} \leq \hat{s} \leq 0$, both satisfy the necessary conditions (4.2)-(4.3). On using (4.3) in (3.69) we find, as asserted, that:

$$E(\hat{s}, \hat{\Delta}) = p[\frac{\mu_b}{\mu_a} L - \hat{s}] \leq p[\frac{\mu_b}{\mu_a} L - \tilde{s}] = E(\tilde{s}, \tilde{\Delta}). \tag{4.23}$$

Next assume $M(\Delta_1) \leq 0$. Let $\hat{s} \geq \tilde{s} \geq 0$ both satisfy the necessary optimality conditions. By a similar substitution as above, we find:

$$E(\hat{s}, \hat{\Delta}) = -p\hat{s} + (p+s)\bar{H}(\hat{s}) + p\frac{\mu_b}{\mu_a} L$$

$$\leq -p\tilde{s} + (p+h)\bar{H}(\tilde{s}) + p\frac{\mu_b}{\mu_a} L = E(\tilde{s}, \tilde{\Delta}). \tag{4.24}$$

The inequality follows from the observation that $J(s) \equiv -ps + (p+h)\bar{H}(L, s)$, where $s \equiv s^o(\Delta)$ is decreasing in s as $dJ/ds = -p + (p+h)H(L, s) \leq 0$ by Lemma 3.

Our next concern is with the unimodality of $E(s, \Delta)$. Because of convexity in s for every $\Delta \geq 0$, it is sufficient to consider the shape of $E(s^o(\Delta), \Delta)$ on $\Delta \geq 0$. It turns out that the cost rate function is unimodal if and only if $s^o(\Delta)$ is monotone decreasing in Δ or, equivalently, if and only if $\bar{F}(0)$ is monotone increasing in Δ.

Theorem 4.3 *The following statements are equivalent on $\Delta \geq 0$:*

1. $E(s^o(\Delta), \Delta)$ is unimodal (pseudoconvex) for every $p \geq 0$, $h \geq 0$ $(p + h > 0)$, and $k \geq 0$;

2. $s^o(\Delta)$ is monotone decreasing in Δ;

3. $1 - F(0)$ is monotone increasing in Δ.

Proof We have from (4.2) and (3.53) that:

$$\frac{ds^o(\Delta)}{d\Delta} = -\frac{\partial \bar{F}(0)}{\partial \Delta} \Big/ \frac{\partial \bar{F}(0)}{\partial s} \tag{4.25}$$

Since $\partial \bar{F}(0)/\partial s \geq 0$, we conclude that 2. and 3. are equivalent.

To complete the proof, we may regard the left-hand-side, $\Phi(\Delta)$, of (4.3) a function of Δ only, with $s \equiv s^o(\Delta)$ determined from (4.2). Since $\Phi(0) = 0$, it is clear that $E(s^o(\Delta), \Delta)$ is unimodal for every p, h and k if and only if $\Phi(\Delta)$ is monotone increasing on $\Delta \geq 0$. $d/d\Delta \, \Phi(\Delta) > 0$ can be written, after substituting (4.2) as:

$$\frac{d}{d\Delta}\Phi(\Delta) > 0 \Leftrightarrow \begin{cases} s'[\dfrac{p}{p+h} - H(L,s)] < 0 & , \text{ if } s \geq 0, \\[2mm] s' < 0 & , \text{ if } s < 0. \end{cases} \tag{4.26}$$

The equivalence of 1. and 2. then follows from Lemma 4.2.

Theorem 4.3 may be useful in many cases to establish the unimodality of $E(s, \Delta)$ through the monotonicity of $\bar{F}(0)$ or of $s^o(\Delta)$. In view of Lemma 3.1, one immediate consequence is the following.

Corollary 4.1 *$E(s, \Delta)$ is unimodal if $1 + R(x)$ is log-concave on $x \geq 0$.*

This result is proved in Stidham (1986) through separate arguments for systems with convex holding and shortage costs. Under
a slightly stronger assumption–that $R(x)$ is concave–a proof that is
more in line with the above developments was given in Şahin (1982).
Unfortunately, $1 + R(x)$ is not concave or log-concave for many distributions of practical interest. The only general class of distributions
that are included are the class of *Decreasing Failure Rate* (DFR)
distributions. It should be emphasized, however, that the results obtained above are not conditional on the specific values of p, h and k.
It is clear that $E(s, \Delta)$ may be convex or pseudoconvex in a given
situation under some values of these parameters without additional
assumptions.

Sharper results can be obtained when $L = 0$ or when batch sizes
are all unity. We now take these up in turn.

4.2   Immediate Deliveries

4.2.1   Continuous Review

For the continuous review model, $L = 0$ implies $D(L) = 0$, $H(0, x) = 1$ and $\bar{H}(0, x) = x$. By specializing the corresponding results obtained
in chapter 3 and in this chapter, we find:

$$1 - F(0) = \begin{cases} 1, & s \geq 0, \\ \dfrac{1 + R(S)}{1 + R(\Delta)}, & s \leq 0, \end{cases} \tag{4.27}$$

and

$$E(s, \Delta) = \begin{cases} \dfrac{k - p[\Delta + \bar{R}(\Delta)] + (p + h)[S + \bar{R}(S)]}{1 + R(\Delta)} - ps & , s \leq 0, \\ \dfrac{k + h[\Delta + \bar{R}(\Delta)]}{1 + R(\Delta)} + hs & , s \geq 0. \end{cases} \tag{4.28}$$

We also observe that $I(\Delta) = 1$ (cf. equation (4.5)) and the necessary optimality conditions, from (4.2)-(4.3), are :

$$\frac{1 + R(S)}{1 + R(\Delta)} = \frac{p}{h + p},\qquad(4.29)$$

and

$$p[\Delta + \bar{R}(\Delta)] - (p + h)[S + \bar{R}(S)] = k.\qquad(4.30)$$

The following are easy consequences of these and the related results proved above.

Corollary 4.2 $s^o(\Delta) \leq 0$ for $\Delta \geq 0$ and $s^* \leq 0$.

Corollary 4.3 $1 - F(0)$ is monotone increasing in Δ, $s^o(\Delta)$ is monotone decreasing in Δ, and $E(s^o(\Delta), \Delta)$ is unimodal for all $p \geq 0$, $h \geq 0$ $(p + h > 0)$, and $k \geq 0$ if and only if $1 + R(x)$ is log-concave.

Corollary 4.2 follows from Lemma 4.1 and the fact that $I(\Delta) = 1$ for $L = 0$. In Corollary 4.3, the statement that $1 - F(0)$ is monotone increasing if and only if $1 + R(x)$ is log-concave follows from (4.27). The rest are then implied by Theorem 4.3. Log-concavity of $1 + R(x)$ was first shown to be both necessary and sufficient in Stidham (1977) for the unimodality of $E(s, \Delta)$ in the case of zero lead times. This condition is sufficient but not necessary for $L > 0$.

4.2.2 Periodic Review

For the periodic review system under immediate deliveries, results of section 4.2 on the fill rate and the cost rate function specialize to $L = 0$ as follows.

$$1 - F(0) = \begin{cases} \dfrac{R(\Delta) + V_\Delta(s)}{1 + R(\Delta)} & , s \geq 0, \\[3mm] \dfrac{R(S)}{1 + R(\Delta)} & , s < 0, \end{cases}\qquad(4.31)$$

$$E(s, \Delta) = \frac{N(s, \Delta)}{1 + R(\Delta)} + p\mu_b, \tag{4.32}$$

where

$$N(s, \Delta) = \begin{cases} K + (h + p)\bar{V}_\Delta(s) + h[sR(\Delta) + \bar{R}(\Delta)] - pS & \text{,if } s \geq 0, \\ \\ K + (h + p)\bar{R}(S) - p[sR(\Delta) + \bar{R}(\Delta) + S] & \text{,if } s \leq 0. \end{cases} \tag{4.33}$$

The optimality conditions, $\partial E/\partial s = 0$ and $\partial E/\partial \Delta = 0$, turn out to be, if $s \geq 0$:

$$\bar{V}_\Delta(s) - \frac{p - hR(\Delta)}{p + h} = 0, \tag{4.34}$$

$$(p + h)\bar{B}(s) - ps - \frac{N(s, \Delta)}{[1 + R(\Delta)]} = 0, \tag{4.35}$$

and, if $s \leq 0$:

$$\frac{R(S)}{1 + R(\Delta)} - \frac{p}{p + h} = 0, \tag{4.36}$$

$$\Delta + \bar{R}(\Delta) - \frac{p + h}{p}\bar{R}(S) - \frac{K}{p} = 0. \tag{4.37}$$

Recall that $V_t(x)$ is the c.d.f. of the forward recurrence time at time t, denoted by V_t, of the renewal sequence of periodic demands (cf. Section 2.1.3). Therefore, V_Δ represents the excess or the *overshoot* with respect to level Δ caused by a period demand that triggers an order. Expressions for $V_t(x)$ and its p.d.f. $v_t(x)$ are given in (2.25) and (2.19)-(2.36).

We next show that $E(s, \Delta)$ is unimodal under a weaker condition than the log-concavity of $1 - R(x)$.

Theorem 4.4 $1 - F(0)$ *is increasing in* Δ *if* $1 + R(x)$ *is log-concave or if* $r(x)$ *is concave increasing.*

Proof That $1 - F(0)$ is increasing in Δ if $1 + R(x)$ is log-concave follows from Lemma 3.1. Assume therefore that $r(x)$ is concave increasing.

For $s \geq 0$, we have:

$$\frac{d}{d\Delta}[1 - F(0)] \geq 0 \Leftrightarrow$$

$$[r(\Delta) + \frac{d}{d\Delta}V_\Delta(s)][1 + R(\Delta)] - r(\Delta)[R(\Delta) - V_\Delta(s)] \geq 0$$

$$\Leftarrow r(\Delta)[1 - V_\Delta(s)] + \frac{d}{d\Delta}V_\Delta(s)[1 + R(\Delta)] \geq 0. \qquad (4.38)$$

The last inequality holds by Lemma 2.3.

For $s \leq 0$, we have:

$$\frac{d}{d\Delta}[1 - F(0)] \geq \Leftrightarrow r(S)[1 + R(\Delta) - r(\Delta)R(S)] \geq 0$$

$$\Leftrightarrow \frac{1 + R(\Delta)}{r(\Delta)} \geq \frac{R(S)}{r(S)}, \quad \Delta \geq S > 0. \qquad (4.39)$$

The last inequality holds if $R(x)/r(x)$ is increasing in $x \geq 0$. This is assured by Lemma 2.4.

As a consequence of Theorems 4.3 and 4.4, we can now state the following.

Corollary 4.4 $s^o(\Delta)$ *is monotone decreasing in Δ and $E(s_o(\Delta), \Delta)$ is unimodal (pseudoconvex) on $\Delta \geq 0$ for all $p \geq 0$, $h \geq 0$ $(p + h > 0)$ and $K \geq 0$, if $1 + R(x)$ is log-concave or $r(x)$ is concave increasing.*

Accordingly, contrary to continuous review systems with zero lead-times, log-concavity of $1 + R(x)$ is not necessary for the unimodality of the cost rate function in periodic review systems with zero lead times. The same is also true for continuous and periodic review systems with positive lead times.

As a simple example, consider the demand density $b(x) = \lambda^2 x e^{-\lambda x}$, $x \geq 0$. The renewal density and the renewal function are given in Section 2.1.4. It can be seen that $r(x)$ is concave increasing and Theorem 4.4 and its corollary apply. But $1 + R(x)$ is not concave or log-concave on $x \geq 0$.

4.3   Unit Demands

For the model of section 3.1.4 under unit demands, we have $1 + R(\Delta) = 1 + \Delta$ which is concave, therefore log-concave. Accordingly, the requirement of Lemma 3.1 is formally satisfied. However, separate proofs are required for the corresponding results on the shapes of $1 - F(0)$ and $E(s, \Delta)$. It is easy to see that $1 - F(0)$ is monotone increasing in Δ. For on using (3.81) we have:

$$\frac{1}{\Delta + 2} \sum_{j=[s-1]+}^{S} H(j) - \frac{1}{\Delta + 1} \sum_{j=[s-1]+}^{S-1} H(j)$$

$$= \frac{H(S)}{\Delta + 2} - \frac{1}{(\Delta + 2)(\Delta + 1)} \sum_{j=[s-1]+}^{S-1} H(j) > \frac{1}{\Delta + 2}[H(S) - H(S-1)].$$

(4.40)

The inequality follows from the facts that $H(j) < H(S-1), j < S-1$, and $\sum_{j=[s-1]+}^{S-1} = S \leq \Delta + 1$ if $s \leq 0$, $= \Delta + 1$ if $s > 0$.

This finding anticipates that $E(s, \Delta)$ is unimodal without additional assumptions. This in fact holds as we next show.

Theorem 4.5 *In order for (s^*, Δ^*), $s^* + \Delta^* = S^*$, to be global minimum of $E(s, \Delta)$ it is necessary and sufficient that*

$$\frac{1}{1 + \Delta^*} \sum_{n=[s^*]+}^{S^*} H(n) > \frac{p}{h + p} \geq \frac{1}{1 + \Delta^*} \sum_{n=[s^*-1]+}^{S^*-1} H(n), \qquad (4.41)$$

and

$$\frac{2}{(2+\Delta^*)(1+\Delta^*)} \left[\sum_{n=[s^*]+}^{S^*} (n - s^* + 1)H(L, n) - \frac{k}{p + h} \right] > \frac{p}{h + p} \geq$$

$$\frac{2}{\Delta^*(1+\Delta^*)} \left[\sum_{n=[s^*]+}^{S^*-1} (n - s^* + 1)H(L, n) - \frac{k}{p + h} \right].$$

(4.42)

Proof For the first order differences, we have, after some routine work:

$$D_s E(s, \Delta) = E(s+1, \Delta) - E(s, \Delta)$$

$$= -p + \frac{h+p}{1+\Delta} \sum_{n=[s]+}^{S} H(L, n) \qquad (4.43)$$

$$D_\Delta E(s, \Delta) = E(s, \Delta+1) - E(s, \Delta)$$

$$= \frac{1}{(\Delta+1)(\Delta+2)} [(h+p) \sum_{n=[s]+}^{S} (n-s+1)H(L, n) - k] - \frac{p}{2}. \quad (4.44)$$

Similarly, for the second order differences we find:

$$D_s E(s, \Delta) - D_s E(s-1, \Delta)$$

$$= \frac{h+p}{1+\Delta} [\sum_{n=[s]+}^{S} H(L, n) - \sum_{n=[s-1]+}^{S-1} H(L, n)] > 0, \qquad (4.45)$$

$$D_\Delta E(s, \Delta) - D_\Delta E(s, \Delta-1) = \frac{(h+p)H(L, S)}{\Delta+2}$$

$$+ \frac{2}{\Delta(\Delta+1)(\Delta+2)} [k - (h+p) \sum_{n=[s]+}^{S-1} (n-s+1)H(L, n)]$$

$$> \frac{2k}{\Delta(\Delta+1)(\Delta+2)} + \frac{h+p}{\Delta+2} [H(L, S) - H(L, S-1)] > 0, \quad (4.46)$$

where we used the inequality

$$\sum_{n=[s]+}^{S-1} (n-s+1)H(L, n) < H(L, S-1) \sum_{n=s}^{S-1} (n-s+1)$$

$$= H(L, S-1) \frac{\Delta(\Delta+1)}{2}. \qquad (4.47)$$

In view of (4.45) and (4.46), we conclude that $E(s, \Delta)$ is pointwise convex. Therefore, the necessary optimality conditions are also sufficient. These conditions are: $D_\Delta E(s, \Delta) \geq 0$, $D_\Delta E(s, \Delta-1) < 0$,

$D_s E(s, \Delta) \geq 0$, and $D_s E(s-1, \Delta) < 0$; they can be written as (4.41) and (4.42). This completes the proof.

Finally, we note the counterpart of Lemma 4.1 which is useful in computational work.

Corollary 4.5 $s^* \leq 0$ *if and only if* $\Delta^* \geq \Delta_1$, *where* Δ_1 *is defined as the smallest non-negative integer for which*

$$\varphi(\Delta_1) \equiv \frac{1}{1 + \Delta_1} \sum_{n=0}^{\Delta_1} H(L, n) > \frac{p}{p + h}. \qquad (4.48)$$

Proof Let $\Delta \geq 0$ be arbitrary but fixed. From convexity, it follows that the reorder point s that minimizes $E(s, \Delta)$ is negative if and only if $D_s E(s, \Delta)|_{s=0} > 0$. This amounts to $\varphi(\Delta) > p/(p + h)$. The corollary follows from the fact that $\varphi(\Delta)$ is monotone increasing in Δ as:

$$\varphi(\Delta + 1) - \varphi(\Delta) = \frac{1}{2 + \Delta}[H(L, \Delta + 1) - \frac{1}{1 + \Delta} \sum_{n=0}^{\Delta} H(L, n)]$$

$$> \frac{1}{2 + \Delta}[H(L, \Delta + 1) - H(L, \Delta)] > 0. \qquad (4.49)$$

In particular, in the case of immediate deliveries, we have $H(L, n) = 1$, $n \geq 0$; therefore $\varphi(\Delta) = 1$ for $\Delta \geq 0$, $s^* \leq 0$.

Chapter 5

Optimal Policies and Approximations

For the inventory systems investigated in the previous chapters, the stationary optimal policy can be determined in principle by finding the global minimum of the corresponding cost rate function. (For previous work in this area, see, for example, Veinott and Wagner, 1965; Archibald and Silver, 1978; and Federgruen and Zipkin, 1984). However, computational difficulties make the exact models unattractive in practice. This has motivated several approaches for the determination of approximately optimal (s, S) policies that require less computational effort and demand information. For continuous review systems, most of the earlier work was confined to the unit demands case under an order-quantity, reorder-point policy. This literature is based on approximations derived by Hadley and Whitin (1963) and Wagner (1969) which require the simultaneous iterative solution of two equations (cf. Gross and Ince, 1975; Gross, Harris and Roberts, 1972; Nahmias, 1976; and others). For periodic review systems, approximately optimal policies that require little computational effort were proposed previously by Naddor (1975, 1980), Ehrhardt (1979), Ehrhardt and Kastner (1980), Ehrhardt and Mosier (1984), Freeland and Porteus (1980), Federgruen and Zipkin (1983), and others. These and other approaches have been reviewed by Porteus (1985).

Although reduced significantly, the computational effort required for the implementation *in practice* of most of these approximations is still discouraging. One exception is the Power Approximation for periodic review systems (Ehrhardt, 1979; Ehrhardt and Mosier, 1984) which is distribution-free and easily computable. Its accuracy is questionable, however, under some parameter configurations (see part 5.1.3 below).

In order to be widely usable in practice, any approximation must fulfill two general requirements: 1) it must be accurate for *pre-specified* and *wide range* of parameter values, and 2) it must be *distribution-free* and *easily computable*. In relation to the first requirement, it should also be emphasized that by accurate we mean practically the same as the optimal policy. As in most cases the cost rate function is rather *flat* around its global minimum, an approximation error that results in more than one or two percent relative cost increase should be considered excessive. It is important that the range of parameter values for which such accuracy is achieved by the approximation is also specified as an integral part of the approximation. Clearly, any approximation will perform well under some scenarios and fail under others.

In view of these considerations, perhaps the most promising approach to developing accurate approximations to stationary control policies that requires minimal computational effort and demand information is the one based on asymptotic renewal theory (cf. Roberts, 1962; Ehrhardt, 1979; Ehrhardt and Mosier, 1984; Tijms and Gronevelt, 1984; Sahin and Sinha,1987). In this chapter, we will review some of these approximations. Following a brief look at the unit demands case, our primary concern in section 5.1 will be with the identification of simple, distribution-free conditions under which asymptotic approximations are accurate for continuous and periodic review systems with zero lead times.

Under these conditions, we will compare the performances of various approximations with those of the optimal policies using gamma, Weibull, truncated normal, inverse Gaussian, and log-normal distributions for demand. In section 5.2, we will extend these approximations to positive lead times under the additional simplification that the lead time demand distribution is approximately normal.

5.1 Immediate Deliveries

For $L = 0$ the models and computational problems are much simpler. The case is of practical interest when the lead time is negligible. We begin with unit demands, followed by batch demands and periodic review systems.

5.1.1 Unit Demands

Recall from section 4.3 that the cost rate function is pointwise convex and that $s^* < 0$ in the case of immediate deliveries. For $L = 0$, the optimality conditions (4.41)-(4.42) are reduced to:

$$1 + S^* > \frac{p[1 + \Delta^*]}{p + h} \geq S^*, \qquad (5.1)$$

$$2 + \Delta^* > \frac{2k - (p + h)s^*(1 - s^*)}{h(1 + \Delta^*)} \geq \Delta^* \qquad (5.2)$$

Thus the optimal policy is distribution-free. (Actually it only depends on the mean interdemand time through $k = K/\mu_a$.) And, it can be very easily computed by a simple algorithm. Some examples are provided by Table 5.1 where we parametrized by:

$$p/h: \quad 1, \quad 2, \quad 3, \quad 5, \quad 10, \quad 20$$
$$2K/\mu_a h: \quad 1, \quad 2, \quad 3, \quad 5, \quad 10, \quad 20, \quad 50, \quad 100, \quad 200$$

Note that the order size is $\Delta^* + 1$ and the reorder point is $s^* - 1$, due to our convention of ordering up to S^* when the inventory position is $< s^* (= s^* - 1)$.

Table 5.1

Optimal Policies Under Unit Demands and Immediate Deliveries

$2K/\mu_a h$	p/h	s^*	Δ^*
1, 2	all	0	0
3, 4	all	0	1
10	1	-1	3
	≥ 2	0	2
20	1	-2	5
	2	-1	4
	≥ 3	0	3
50	1	-4	9
	2	-2	7
	3	-1	7
	5	-1	7
	10	0	6
	20	0	6
100	1	-6	13
	2	-3	11
	3	-2	10
	5	-1	10
	10	0	9
	20	0	9
200	1	-9	19
	2	-5	16
	3	-3	15
	5	-2	14
	10	-1	14
	20	0	13

5.1.2   Batch Demands

For the continuous review system with batch demands, the cost rate function is given by (4.28) and the optimality conditions by (4.29)-(4.30). We also know in this case that $s^* \leq 0$ (Corollary 4.2) and if there are multiple solutions to the optimality conditions, the one with the largest s represents the global minimum (Theorem 4.2), subject to condition (4.7).

When it holds, the last result reduces the optimal policy computations to a one-dimensional search process. Starting from a suitable initial $s^{(0)} < 0$, for example, trial values of s can be generated by $s^{(n)} = s^{(n-1)} - \epsilon_n$ where $\epsilon_n > 0$ is a suitable step size. At each iteration $\Delta^{(n)}$ can be computed from (4.29). If (4.30) also holds, then $(s^{(n)}, \Delta^{(n)})$ is optimal. Note that for a given $s^{(n)}$, $\Delta^{(n)}$, computed from (4.29) or (4.18), may not be unique. But, in any event, the $\Delta^{(n)}$ value that together with $s^{(n)}$, also satisfies the other condition is optimal. According to Corollary 4.3, the optimality conditions are also sufficient if the renewal function of the batch size distribution is log-concave.

We have used this algorithm successfully with the gamma, Weibull, truncated normal, inverse Gaussian, and lognormal batch size distributions. As noted at the beginning of Chapter 4, $E(s, \Delta)$ is convex in s for $\Delta \geq 0$ fixed. For fixed s, on the other hand, $E(s, \Delta)$ assumed different shapes as a function of Δ and of the other parameter values. Three of the most frequently encountered shapes are shown in Figures 5.1, 5.2, and 5.3 for $s = 0$. The convex shape seen in Figure 5.1 is typical of DFR gamma and Weibull distributions (for which $R(x)$ is concave), irrespective of other parameter values, and of all distributions for k large enough. What is large enough depended, however, on the distribution and the shape parameter value. For the gamma distribution, for example, the shape in Figure 5.1 prevailed for $\alpha = 2$ even when $k = 10$. For $\alpha = 7$, however, the shape of $E(0, \Delta)$ was of the Figure 5.2 type all the way to $k = 100$. Generally speaking, high coefficients of variation (> 1) induced the Figure 5.1 shape, lower dispersion ($0.3 < c_b < 1$) resulted in shapes like in Fig-

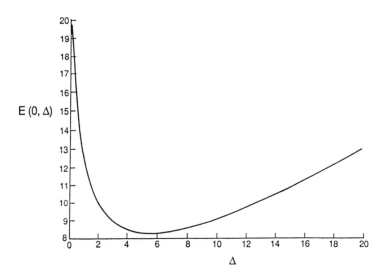

Fig.5.1. The cost rate fuction $(s = 0)$ for $k = 20$ under log-
normal batch sizes with $\alpha = 1.5$

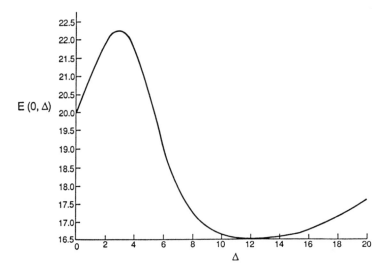

Fig.5.2. The cost rate function $(s = 0)$ for $k = 20$ under
gamma batch sizes with $\alpha = 20$

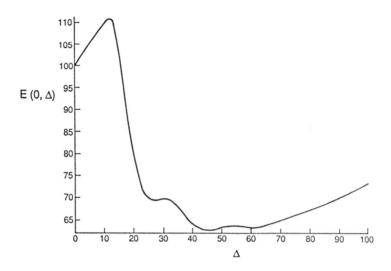

Fig.5.3. The cost rate function ($s = 0$) for $k = 100$ under inverse Gaussian batch sizes with $\alpha = 20$

ure 5.2 and lowest values of $c_b(< 0.3)$ were associated with the shape seen in Figure 5.3. Higher values of k ($\equiv K/\mu_a$) had a moderating effect on the shape of the cost rate function (i.e., Figure 5.2 rather than 5.3 or 5.1 rather than 5.2).

These observations are based on a scale parameter value of $\beta = 1$ in all distributions. They remain valid, however, for other values of β. Values of the cost rate function and the control parameters under any scale parameter β can be related to their corresponding values under $\beta = 1$. For this we refer to the relations noted at the end of section 2.2.1. (As in that section, we modify the notation in an obvious way to emphasize certain relationships.) Since $R(\beta; x) = R(1; x/\beta)$ and $\bar{R}(\beta; x) = \beta \bar{R}(1; x/\beta)$, we find from (4.28) that:

$$E(\beta; s, \Delta, k) = \beta E(1; s/\beta, \Delta/\beta, k/\beta). \tag{5.3}$$

It follows that $s^*(\beta; k) = \beta s^*(1; k/\beta)$, $\Delta^*(\beta; k) = \beta \Delta^*(1; k/\beta)$, and $E^*(\beta; k) = \beta E^*(1; k/\beta)$. Thus the optimal values of both control

parameters and the cost rate function under an arbitrary scale parameter β and setup cost K are simply β times the corresponding
quantities under $\beta = 1$ and setup cost K/β.

The cost rate function (4.28) ($s \leq 0$) can be approximated by
replacing, for S large enough, $\bar{R}(\Delta)$, $\bar{R}(S)$ and $R(\Delta)$ by their asymptotic approximations (cf. (2.70)):

$$R(t) \simeq \frac{t}{\mu_b} + \frac{c_b^2 - 1}{2}, \tag{5.4}$$

$$\bar{R}(t) \simeq \frac{t^2}{2\mu_b} + \frac{c_b^2 - 1}{2}t. \tag{5.5}$$

It is seen that the approximate cost rate function is minimized at

$$s^a = -\mu_b h \sqrt{\frac{1}{p(p + h)}[\frac{2k}{h\mu_b} - C_b^2]} \tag{5.6}$$

$$\Delta^a = -\mu_b C_b + \mu_b \sqrt{(1 + \frac{h}{p})[\frac{2k}{h\mu_b} - C_b^2]} \tag{5.7}$$

where $C_b = (1 + c_b^2)/2$ and $k = K/\mu_a$ as before. The implied order-
up-to level is:

$$S^a = -\mu_b C_b + \mu_b \sqrt{(\frac{p}{p + h})[\frac{2k}{h\mu_b} - C_b^2]} \tag{5.8}$$

According to our findings in part 2.2.3, these approximations would
be accurate for the class of distributions under consideration (and
for others with similar characteristics) if $S^a \geq \rho(\epsilon)\mu_b$ for a suitable
$\epsilon > 0$. Note that since $\Delta^a > S^a$, if S^a is large enough in this sense,
so is Δ^a; and S^a, Δ^a, and s^a are all accurate. The most suitable
value for ϵ among the set $\{0.01, 0.05, 0.10, 0.20\}$ turned out to be
0.20. This selection was based on extensive comparative numerical
work; it represents a conservative choice (cf. Sinha, 1985). Using the
approach discussed in part 2.2.3, the convergence measure $\rho(0.20)$
was then estimated to be:

$$\rho(0.20) = 1.615c_b^2 - 2.397c_b + 1.353. \tag{5.9}$$

Thus, if

$$S^a \geq \rho(.20)\mu_b, \tag{5.10}$$

then the approximations (5.6), (5.7) and (5.8) are expected to be accurate. Evidently, both the approximation and the accuracy condition are distribution free; they depend on the mean interdemand time μ_a ($k = K/\mu_a$) and the first two moments of the batch size distribution.

If (s^a, Δ^a) are accurate in the above sense then $-ps^a$ is a good approximation for $E(s^*, \Delta^*) = -ps^*$ of the zero lead time model (cf. (4.10)). Since $E(0,0) = k$, the *stay-out-of-business* condition can be approximated by:

$$s^a \leq -k/p \tag{5.11}$$

Thus, if (5.11) holds, then $s^a = \Delta^a = 0$. However, this should be contingent on condition (5.10), for otherwise s^a may not be sufficiently accurate. Based on these, an algorithm for asymptotic approximations can be described as follows:

1. *Compute S^a from (5.8).*

2. *If (5.10) does not hold, then the approximations may not be accurate.*

3. *If (5.10) holds, compute s^a from (5.6).*

4. *If $s^a \leq -k/p$, then take $s^a = \Delta^a = 0$; the item need not be stocked.*

5. *If $s^a > -k/p$, then s^a and Δ^a, computed (5.7), will be accurate.*

Performance of the approximate control policy (s^a, Δ^a) has been tested extensively, using the demand batch size distributions mentioned above. Some results for $p \to \infty$ are presented in Table 5.2.

Note, from part 4.2.1 that as $p \to \infty$, $s^* \to 0$, and Δ^* is obtained

Table 5.2

Performance of Asymptotic Approximations for Continuous Review Systems ($L = 0$).

Dist.	α	c_b	$K/\mu_a h$	Δ^a	Max. RE (%)
GA	7.0	0.38	≤ 6	n.a.	
			7-12	0.00	0.00
			≥ 13	Δ^*	0.00
	4.0	0.50	≤ 3	n.a.	
			4-7	0.00	0.00
			≥ 8	Δ^*	0.01
	3.0	0.58	≤ 2	n.a	
			3-5	0.00	0.00
			≥ 6	Δ^*	0.01
	2.0	0.71	≤ 2	n.a.	
			3	0.00	0.00
			≥ 4	Δ^*	0.00
	1.5	0.82	1	n.a.	
			2	0.00	0.00
			≥ 3	Δ^*	0.01
	0.6	1.29	≤ 2	n.a.	
			≥ 3	Δ^*	0.03
WL	7.0	0.17	1	n.a	
			≥ 2	Δ^*	9.18†
	4.0	0.28	1	0.00	0.00
			≥ 2	Δ^*	0.94
	3.0	0.36	1	0.00	0.00
			≥ 2	Δ^*	0.01
	2.0	0.52	1	0.00	0.00
			≥ 2		0.11
	1.5	0.68	1	0.00	0.00
			≥ 2	Δ^*	0.01
	0.6	1.76	≤ 16	n.a.	
			≥ 17	Δ^*	0.41#
TN	4.0	0.25	≤ 4	n.a.	
			5-7	0.00	0.00
			≥ 8	Δ^*	3.74#
	2.0	0.46	1	n.a.	
			2-3	0.00	0.00
			≥ 4	Δ^*	0.06
	1.0	0.62	1	n.a.	
			2	0.00	0.00
			≥ 3	Δ^*	0.12
	0.0	0.76	1	0.00	0.00
			≥ 2	Δ^*	0.01
	-1.0	0.85	≥ 1	Δ^*	0.42
	-2.0	0.91	≥ 1	Δ^*	0.05

Table 5.2 (continued)

Dist.	α	c_b	$K/\mu_a h$	Δ^a	Max. RE (%)
IG	20.0	0.22	23	n.a.	
			24-36	0.00	0.00
			$\geq$ 37	Δ^*	5.79#
	12.0	0.29	12	n.a.	
			13-22	0.00	0.00
			$\geq$ 23	Δ^*	0.03#
	8.0	0.35	$\leq$ 7	n.a.	
			8-14	0.00	0.00
			$\geq$ 15	Δ^*	0.01#
	4.0	0.50	$\leq$ 3	n.a.	
			4-7	0.00	0.00
			8	Δ^*	0.01
	1.0	1.0	1	n.a.	
			2	Δ^*	0.94
	0.5	1.41	$\leq$ 2	n.a.	
			3	Δ^*	0.94
LN	0.1	0.32	1	n.a.	
			$\geq$ 2	Δ^*	0.49
	0.5	0.81	1	n.a.	
			2	0.00	0.00
			$\geq$ 3	Δ^*	0.11
	1.0	1.31	$\leq$ 6	n.a.	
			$\geq$ 7	Δ^*	0.78#
	1.5	1.87	$\leq$ 29	n.a.	
			$\geq$ 30	Δ^*	0.76#
	2.0	2.53	$\leq$ 136	n.a.	
			$\geq$ 137	Δ^*	1.01#
	3.0	4.37	$\leq$ 2484	n.a.	

[†]High relative errors caused by small cost rate function values (see Appendix 1)

[#]Maximum error given is the next value of $K/\mu_a h$ from the set $\{1, 2, \ 3, 5, 10, 20, 50, 100, 200\}$ (see Appendix 1).

by minimizing:

$$E(0, \Delta) = \frac{k + h[\Delta + \bar{R}(\Delta)]}{1 + R(\Delta)} \tag{5.12}$$

so that the optimization problem is one-dimensional. The same problem arises by restricting s to non-negative values so that $s^* = 0$ when $L = 0$ (cf. Şahin and Sinha, 1987). It can be seen that the asymptotic approximation for Δ^* in this case is:

$$\Delta_{(s=0)}^a = -\mu_b C_b + \mu_b \sqrt{\frac{2k}{h\mu_b} - C_b^2} \tag{5.13}$$

and that $\lim_{p \to \infty} S^a = \lim_{p \to \infty} \Delta^a = \Delta_{(s=0)}^a$, as expected. Table 5.2 is based on $\Delta_{(s=0)}^*$, $\Delta_{(s=0)}^a$, the accuracy condition:

$$\Delta_{(s=0)}^a \geq \rho(0.20)\mu_b, \tag{5.14}$$

which corresponds to (5.10), and the do not stock condition:

$$k/h \leq \Delta_{(s=0)}^a + \mu_b C_b. \tag{5.15}$$

This latter condition is obtained from the optimality condition $\partial/\partial\Delta$ $E(0, \Delta) = 0$ which implies:

$$E(\Delta^*) = \frac{h[1 + R(\Delta^*)]}{r(\Delta^*)} \simeq h[\Delta^* + \mu_b C_b], \tag{5.16}$$

where we used the asymptotic approximations for R and r.

The measure used in assessing the performances of the various approximations was the relative cost increase due to approximation, computed from:

$$RE = 100\frac{E(s^a, \Delta^a) - E(s^*, \Delta^*)}{E(s^*, \Delta^*)} \tag{5.17}$$

We noted above that if β is a scale parameter of the batch size distribution $B(x)$, then (with an appropriate modification in our notation) $E^*(\beta; k) = \beta E^*(1; k/\beta)$. Similarly, it can be seen that $E^a(\beta; k) = \beta E^a(1; k/\beta)$. This follows from $\Delta^a(\beta; k) = \beta\Delta^a(1; k/\beta)$ and $s^a(\beta; k) = \beta s^a(1; k/\beta)$ which in turn implied by (5.7) or (5.13) and (5.16), respectively. Consequently, $RE(\beta; k) = RE(1; k/\beta)$ and the performance of $(\Delta^a(\beta; k), s^a(\beta; k))$ is the same as the performance of $(\Delta^a(1; k/\beta), s^a(1; k/\beta))$.

Cases for which Δ^a is imaginary, or the accuracy condition (5.14) is not satisfied are marked n.a. in Table 5.2. For all other parameter values, Δ^a is *large enough* to be a good approximation for Δ^*. These include the cases for which $\Delta^a = 0$, as induced by condition (5.15). Because of the use of asymptotic approximations in its construction, (5.15) should be applicable in approximating the do not stock condition essentially when Δ^a is large enough in the sense of condition (5.14). However, condition (5.15) was generally applicable in predicting $\Delta^* = 0$, even when (5.14) did not hold. Also, for most of the cases examined $\Delta^* = 0$ when Δ^a was negative or imaginary. In spite of these, Table 5.2 is filled conservatively by accepting Δ^a as accurate only on the basis of (5.14).

The important conclusion from Table 5.2 is that $\Delta^a_{(s=0)}$ is applicable and accurate under most parameter configurations of practical interest. Applicability of the approximation deteriorates quickly, however, when $c_b > 1.5$. When condition (5.14) holds, the approximations are practically the same as the optimal policies. Results for a set of values of $K/\mu_a h$ are presented in more detail in Appendix 1. These include the exact and approximate policies, and the corresponding approximation errors.

The degree of accuracy of the approximations computed from (5.6)-(5.7) for finite p should be the same as those for $p \to \infty$, in the sense that whenever the accuracy condition (5.10) holds (s^a, Δ^a) should be practically the same as (s^*, Δ^*). This observation is based on the extensive numerical work performed on the periodic review model, as reported in the next part. However, the range of applicability of the approximations will be reduced in that condition (5.10) is a stronger condition than (5.14). This is clear from (5.8) because S^a is increasing in ρ. In spite of this negative effect, condition (5.10) is still weak enough to hold for most parameter configurations of practical interest. Table 5.3 presents these configurations in terms of the values of c_b, p/h and the lower bound on $K/h\mu_a\mu_b$ for which condition (5.10) holds. The latter is the average set up cost per unit per unit time, measured in units of h.

5.1.3   Periodic Review

The idea of using renewal-theoretic asymptotic approximations for the optimal stationary periodic review policies goes back to Roberts (1962). In terms of the zero-lead-time model, he showed that *if $K \to \infty$ and $p \to \infty$, in such a way as to keep s^* fixed, then:*

$$\Delta^* = \sqrt{\frac{2\mu_b K}{h}} + o(1). \qquad (5.18)$$

Thus if Δ^* is large enough, then it can be approximated by the *economic order quantity* which requires only the expected demand. Complete demand distribution is still needed, however, to compute an approximation for s^* from:

$$\int_{s^*}^{\infty} (x - s^*)b(x)dx = \frac{h\Delta^*}{h + p} + o(\Delta^*). \qquad (5.19)$$

To eliminate this need, Wagner (1975) modified the approximations by substituting a normal distribution for the actual demand distribution. More recently, Ehrhardt (1979) and Ehrhardt and Mosier (1984) used regression analysis to fit approximations of the above form to a set of known optimal policies.

From a practical point of view, the critical question, as in part 5.1.2, is: how large is large enough for K and p and in what relationship should they grow? These questions were addressed in Şahin and Sinha (1987) through a new set of approximations based on asymptotic renewal theory. In this part we discuss these approximations and their accuracy.

We begin by specializing some of the results we proved in section 4.1 to periodic review systems under immediate deliveries. First,

Table 5.3

Lower Bound (LB) on $K/h\mu_a\mu_b$ for the Accuracy of Asymptotic
Approximations for Continuous Review Systems $(L = 0)$

c_b	p/h	LB	c_b	p/h	LB
.25	1	2.06	1.25	1	5.49
	2	1.58		2	4.32
	3	1.42		3	3.94
	5	1.29		5	3.62
	10	1.20		10	3.39
	20	1.15		20	3.27
	50	1.12		50	3.20
	100	1.11		100	3.18
	200	1.11		200	3.17
.50	1	1.60	1.50	1	10.42
	2	1.25		2	8.14
	3	1.13		3	7.39
	5	1.04		5	6.78
	10	0.97		10	6.32
	20	0.93		20	6.10
	50	0.91		50	5.96
	100	0.90		100	5.91
	200	0.90		200	5.89
.75	1	1.86	1.75	1	19.16
	2	1.47		2	14.89
	3	1.34		3	13.46
	5	1.24		5	12.32
	10	1.16		10	11.47
	20	1.12		20	11.04
	50	1.10		50	10.78
	100	1.09		100	10.70
	200	1.08		200	10.66
1.00	1	2.97	2.00	1	33.58
	2	2.35		2	25.97
	3	2.15		3	23.43
	5	1.98		5	21.40
	10	1.87		10	19.88
	20	1.80		20	19.12
	50	1.76		50	18.06
	100	1.75		100	18.51
	200	1.74		200	18.43

by Lemma 4.1 and the discussion following its statement, we know that $s^o(\Delta) > (=)0$ if and only if $\Delta < (=)\Delta_1$ where Δ_1 is uniquely determined by $R(\Delta_1) = p/h$. The latter characterization follows from (4.3) which for the periodic review system with $L = 0$ takes the form:

$$\frac{1}{1 + R(\Delta_1)}[B(\Delta_1) + \int_0^{\Delta_1} r(\Delta_1 - u)B(u)du] = \frac{R(\Delta_1)}{1 + R(\Delta_1)} = \frac{p}{p + h}. \tag{5.20}$$

Theorem 4.1 then implies that $\Delta^* > (=)\Delta_1$ and $S^* > (=)\Delta_1$ if and only if:

$$M(\Delta_1) \equiv K - p\Delta_1 + h\bar{R}(\Delta_1) > (=)0. \tag{5.21}$$

Consequently, if $M(\Delta_1) > 0$ then $s^* < 0$ and $\Delta^* > S^* > \Delta_1$, and if $M(\Delta_1) \leq 0$ then $s^* \geq 0$ and $\Delta^* \leq S^* \leq \Delta_1$. We now consider the consequences of these in turn.

For $M(\Delta_1) > 0$, since Δ_1 is a lower bound on S^* and Δ^*, if Δ_1 is *large* then so are S^* and Δ^*. Also, $s^* < 0$ and the optimality conditions are given by (4.36)-(4.37). In terms of the convergence measure introduced above, Δ_1 is large if $\Delta_1 \geq \mu_b\rho(\epsilon)$ for a suitable ϵ. It can then be approximated by using the asymptotic form of $R(t)$ in $R(\Delta_1) = p/h$. This results in

$$\Delta_1 \simeq \mu_b[\frac{p}{h} - \frac{c_b^2 - 1}{2}]. \tag{5.22}$$

Also, on substituting the asymptotic form of $\bar{R}(t)$ in $M(\Delta_1) > 0$, we find:

$$K > \frac{h\Delta_1^2}{2\mu_b}. \tag{5.23}$$

Putting these together with $\Delta_1 \geq \mu_b\rho(\epsilon)$, we obtain:

$$\rho(\epsilon) + C_b \leq 1 + \frac{p}{h} < C_b + \sqrt{\frac{2K}{\mu_b h}}, \tag{5.24}$$

where $C_b = (1 + c_b^2)/2$. The right-hand-side inequality implies $M(\Delta_1) > 0$, so that $s^* < 0$ and $\Delta^* > S^* > \Delta_1$, while the left-hand-side inequality stipulates what constitutes a large enough p to ensure that Δ_1 is large.

In emprical evaluations of the approximations and their accuracy conditions, the left-hand-side condition for large Δ_1 turned out to be too stringent. Note that the lower bound in question decreases with increasing ϵ. However, even for $\epsilon = 0.20$, this condition excluded parameter configurations resulting in highly accurate approximations and larger ϵ values were not consistently appropriate. As an alternative, we used the condition:

$$\frac{p}{h} \geq \rho(\epsilon) \tag{5.25}$$

to identify large p (and Δ_1). This is obtained by approximating $R(\Delta_1)$ by Δ_1/μ_b in $R(\Delta_1) = p/h$.

Under these stipulations, if we replace the renewal function and its integral in the optimality conditions (4.36) and (4.37) by their asymptotic approximations, and solve the resulting system, we obtain the following approximations for (s^*, Δ^*):

$$s^a = \mu_b(1 - \frac{h}{p}A), \tag{5.26}$$

$$\Delta^a = \mu_b[(1 + \frac{h}{p})A - C_b], \tag{5.27}$$

where

$$A = \{\frac{p}{p+h}[\frac{2K}{h\mu_b} + \frac{pc_b^2}{h} - (\frac{c_b^2 - 1}{2})^2]\}^{1/2}. \tag{5.28}$$

These approximations are distribution-free and easily computable.

For $M(\Delta_1) \leq 0$, we have $s^* \geq 0$ and $\Delta_1 \geq S^* \geq \Delta^*$. Therefore, if Δ^* is large, then so are S^* and Δ_1. On using the asymptotic approximations for $R(t)$ and $\bar{R}(t)$ (cf. (5.4) and (5.5)) and the limiting form of the forward recurrence time distribution (cf. 2.60) in the optimality conditions (4.34)-(4.35), we find:

$$\int_0^{s^a} [1 - B(u)]du = \frac{\mu_b(p + h - hC_b) - h\Delta^a}{p + h}, \tag{5.29}$$

and

$$\Delta^a = -(\mu_b C_b + s^a) + \{\frac{2K\mu_b}{h} + \mu_b^2 C_b[2(1 + \frac{p}{h}) - C_b]$$

$$+(s^a)^2 - 2(1 + \frac{p}{h})\tilde{B}(s^a)\}^{1/2}$$

(5.30)

where

$$\tilde{B}(x) = \int_0^x u[1 - B(u)]du.$$

(5.31)

Conditions of validity are:

$$1 + \frac{p}{h} \geq C_b + \sqrt{\frac{2K}{h\mu_b}},$$

(5.32)

which approximates $M(\Delta_1) \leq 0$, and

$$\Delta^a \geq \mu_b\rho(\epsilon).$$

(5.33)

Note, however, that (5.33) is not sufficient for the accuracy of (s^a, Δ^a). It may be that Δ^* is small and Δ^a is inaccurate and large in the sense of (5.33). Choice of an appropriate value for ϵ as well as the validity of these accuracy conditions are based on the empirical considerations that follow.

As we note below, (5.29)-(5.30) can be solved iteratively through a modest computational effort. However, as opposed to the case of $M(\Delta_1) > 0$, determination of (s^a, Δ^a) is not explicit when $M(\Delta_1) < 0$ and the full demand distribution is still required. A simplification can be obtained by noting that:

$$\tilde{B}(s^a) \simeq \frac{s^a}{2} \int_0^{s^a} [1 - B(u)]du$$

$$= \frac{s^a[\mu_b(p + h - hC_b) - h\Delta^a]}{2(p + h)}.$$

(5.34)

This leads to:

$$\Delta^a = -\mu_b C_b - \frac{s^a}{2} + [\mu_b(1 + \frac{p}{h})(2\mu_b C_b - s^a) - \mu_b^2 C_b^2 + \frac{2\mu_b K}{h}]^{1/2}. \quad (5.35)$$

In what follows we also validate (5.35). Its advantage is that *given* s^a, it is distribution-free and easily computable.

The *power approximation* of Ehrhardt (1979) and the *revised power approximation* of Ehrhardt and Mosier (1984) are also based on asymptotic renewal theory. Expressions (5.18) and (5.19) are used to design regression models which are then fitted to a grid of 288 inventory items with known optimal policies, computed under the assumptions of Poisson and negative binomial demand distributions. A fixed lead time of $L \geq 0$ periods is allowed between the placement and delivery of an order. For $L = 0$, the revised power approximation is:

$$s^r = 0.973\mu_b + (1.063 - 2.192z + \frac{0.183}{z})\sigma_b, \qquad (5.36)$$

$$\Delta^r = 1.30\mu_b^{0.494}(1 + C_b^2)^{0.166}(\frac{K}{h})^{0.506}, \qquad (5.37)$$

where $z = [h\Delta^r/p\sigma_b]^{1/2}$.

Using the gamma, Weibull, truncated normal, inverse Gaussian and the lognormal distributions, this time in describing periodic demands, we undertook a comprehensive evaluation of the approximations reported above. As before, it turns out that we could restrict ourselves, without any loss, to a scale parameter value of one in all these distributions. It can be seen, again, that $E(\beta; s, \Delta, K) = \beta E(1; s/\beta, \Delta/\beta, K/\beta)$. Therefore, $s^*(\beta; K) = \beta s^*(1; K/\beta)$, $\Delta^*(\beta; K) = \beta \Delta^*(1; K/\beta)$, and $E^*(\beta; K) = \beta E^*(1; K/\beta)$. These also hold for (s^a, Δ^a), determined by either of (5.26)-(5.27), (5.29)-(5.30), (5.29)-(5.35), and for (s^r, Δ^r). For example, it is seen from (5.26)-(5.27) that $\mu_b(\beta) = \beta\mu_b(1)$ and $c_b(\beta) = c_b(1)$ imply $A(\beta; K) = A(1; K/\beta)$. Therefore, $s^a(\beta; K) = \beta s^a(1; K/\beta)$, $\Delta^a(\beta; K) = \beta \Delta^a(1; K/\beta)$. This ability to restrict β to 1 without losing information made it possible to carry out the computations for wide ranges of the shape parameter values of each of the distributions in the set. The shape parameter values used and the corresponding means ($\beta = 1$) and coefficients of variation were as in Table 2.1. The upper and lower limits on shape parameter values were chosen to include most contingencies that may arise in practice. It is unlikely that outside these limits the distributions in question would arise as demand distributions.

Determination of the optimal policies requires the accurate computation of the renewal function and certain integrals that involve the renewal function. As mentioned previously, for this purpose we used the generalized cubic splining algorithm of McConalogue (1981) which is basically designed to compute convolutions $B_n(x)$ of a distribution. The renewal function is then computed from $R(x) = \sum_1^\infty B_n(x)$, with an appropriate convergence criterion on $B_n(x)$ and other renewal-theoretic functions and terms that appear in the cost rate function and the optimality conditions are computed by integrating the spline representation of $B_n(x)$ or of $R(x)$.

Theorem 4.2 reduces the computation of the optimal policies to a simple search procedure that could be started from a suitable upper bound on s^*. If $M(\Delta_1) \leq 0$, then 0 is an upper bound. If $M(\Delta_1) > 0$, then s^+ is an upper bound, as determined from $B(s^+) = p/(p+h)$; this follows from Lemma 4.2. In the first case we started from the initial approximation $(0, \Delta_1)$, which satisfies (4.36), and in the second case from $(s^+, 0)$, which satisfies (4.34). We then iterated through $s^{(n)} = s^{(n-1)} - \delta_n$ where $\delta_n > 0$ is a suitable step size. At each iteration, given $s^{(n)}$, $\Delta^{(n)}$ was computed from (4.34) or (4.36). If (4.35) or (4.37) also held, $(s^{(n)}, \Delta^{(n)})$ was identified as the optimal policy.

Computation of the power approximation and the asymptotic approximation when $M(\Delta_1) > 0$ are immediate. When $M(\Delta_1) < 0$, the asymptotic approximation can be computed as the simultaneous iterative solution of (5.29)-(5.30) or (5.29)-(5.35) by an algorithm similar to the well-known algorithm of Hadley and Whitin (1963). We may regard (5.29)-(5.30), for example, as equations in s and Δ, with iterative solutions $(s^{(n)}, \Delta^{(n)})$, $n = 0, 1, ...,$ and simultaneous solution (s^a, Δ^a). By iterative solutions we mean $\Delta = \Delta^{(n)}$ is the solution of (5.30) for $s = s^{(n)}$, and $s = s^{(n+1)}$ is the solution of (5.30) for $\Delta = \Delta^{(n)}$, iteratively for $n = 0, 1, ...$. As determined by (5.29), s is monotone decreasing in Δ; and, as determined by (5.30), Δ is monotone decreasing in s. The first observation is evident. The second can be seen from $d\Delta/ds < 0$ which holds provided $\Delta > 0$. Also, for

$s^{(0)} = 0$ we can obtain from (5.30):

$$\Delta^{(0)} = -\mu_b C_b + \{\mu_b^2 C_b[2(1 + p/h) - C_b] + 2K\mu_b/h\}^{1/2} \qquad (5.38)$$

which can be seen to be ≥ 0 under condition (5.32). For $s^{(1)} \geq 0$ we need $(1 + p/h - C_b)\mu_b - \Delta^{(0)} \geq 0$ which also holds under condition (5.32). Consequently, if we start with $s^{(0)} = 0$ and $\Delta^{(0)}$, we have $s^{(1)} \geq 0$. The monotonicity properties noted above then imply that $\{\Delta^{(n)}, \quad n \geq 0\}$ is decreasing and $\{s^{(n)}, \quad n \geq 0\}$ increasing. These sequences converge to (s^a, Δ^a) quite rapidly.

The performance measure used in the evaluation of all approximations was the relative cost increase under the approximate policy, given by (5.17). As in the continuous review case, it can be seen that $RE(\beta; K) = RE(1; K/\beta)$ for all the approximations given above. Therefore, their performances under the demand distribution $B(\beta; x)$ and setup cost K are the same as their respective performances under $B(1; x)$ and K/β.

As noted before, the shape parameter values used in demand distributions were as in Table 2.1. The cost parameter settings were:

$$p/h: \quad 1, \quad 2, \quad 3, \quad 5, \quad 10, \quad 20$$
$$K/h: \quad 5, \quad 10, \quad 20, \quad 50, \quad 100, \quad 200$$

These resulted in 1,080 cases. For each case we computed the optimal policy and the three approximations given above and their costs and relative errors. In terms of the convergence measure $\rho(\epsilon)$, the most appropriate values for ϵ, chosen by trial and error from the set $\{0.01, 0.05, 0.10, 0.20\}$ were $\epsilon = 0.20$ in condition (5.25) and $\epsilon = 0.10$ in condition (5.33). Using the polynomial approximations for $\rho(.20)$ and $\rho(.10)$, introduced in section 2.2.3, an algorithm for asymptotic approximations can be outlined as follows:

1. *If*

$$\frac{p}{h} \geq 1.353 - 2.397c_b + 1.615c_b^2, \qquad (5.39)$$

where c_b is the demand coefficient of variation, go to step 2.; otherwise, (s^a, Δ^a) may not be accurate.

2. If

$$1 + \frac{p}{h} \le C_b + \sqrt{\frac{2K}{h\mu_b}}, \qquad (5.40)$$

where $C_b = (1 + c_b^2)/2$, then $s^a \le 0$ $(s^* \le 0)$ and (s^a, Δ^a), computed from (5.26)-(5.27) will be accurate.

3. If (5.40) holds as >, then $s^a > 0$ $(s^* > 0)$ and (s^a, Δ^a), computed from (5.26)-(5.29) or (5.29)-(5.35) will be accurate, provided that

$$\frac{\Delta^a}{\mu_b} \ge 1.720 - 3.383c_b + 2.479c_b^2; \qquad (5.41)$$

otherwise, (s^a, Δ^a) may not be accurate.

The right-hand-side of (5.39) is $\hat{\rho}(.20)$ and that of (5.41) is $\hat{\rho}(.10)$ (> $\hat{\rho}(.20)$). The reason for the relevance of two different measures of convergence is that (5.39) is a condition for the lower bound Δ_1 of $\Delta^*(\simeq \Delta^a)$ to be large, while (5.41) is a condition for Δ^a itself to be large. Note that in addition to implying accurate (s^a, Δ^a) when (5.40) holds, condition (5.39) is necessary for the accuracy of (5.40). For if (5.39) does not hold (Δ_1 is not large enough) then there is no assurance that (5.40) will correctly predict the sign of s^*. Note also that for $p/h = 1, 2$, and 3 respectively, (5.39) holds when $c_b < 1.32$, 1.72, and 2, and that for $c_b = 0.5, 1.0$, and 1.5, (5.41) holds when $\Delta^a/\mu_b > 0.65, 0.82, 2.22$. Thus both conditions are weak enough to be generally applicable. Regions of accuracy of asymptotic approximations for $c_b = 0.25(0.25)2.00$ are shown in Table 5.4.

Performances of the asymptotic approximation computed with the above algorithm (using (5.29)-(5.30) when (5.40) holds as >) and the power approximation computed from (5.36)-(5.37) are summarized in Table 5.5. In terms of the performance measure RE, both approximations performed well in 936 cases that exclude only the IG distribution with $\beta = 20.0$ and the LN distribution with $\beta \ge 1.5$. These represent (with very few exceptions) cases for which the accuracy conditions are satisfied. Performance of the asymptotic approximation, however, was much better than the power approximation. Detailed results including optimal policies, all three approximations,

Table 5.4

Regions of Accuracy of Asymptotic Approximations for Periodic Review Systems[†] $(L = 0)$

c_b	$\rho(.20)$	$\rho(.10)$	p/h	$(2K/h\mu_b)_{CR}$	p/h	$(2k/h\mu_b)_{CR}$
0.25	0.855	1.029	1	1.49	15	231.6
			3	10.36	30	913.2
			5	27.24	50	2520.9
			10	104.42	100	10043.8
0.50	0.558	0.648	1	0.77	15	221.3
			3	8.27	30	892.5
			5	23.77	50	2487.5
			10	97.52	100	9975 .0
0.75	0.464	0.577	1	0.22	15	209.3
			3	6.10	30	868.4
			5	19.97	50	2447.2
			10	89.66	100	9894.0
1.00	0.571	0.816	1	0.00	15	196.0
			3	4.00	30	841.0
			5	16.00	50	2401.0
			10	81.00	100	9801.0
1.25	0.880	1.365	1	0.28	15	181.4
			3	2.16	30	810.5
			5	12.03	50	2349.2
			10	71.72	100	9696.1
1.50	1.391	2.223	1	1.27	15	165.8
			3	0.77	30	777.0
			5	8.27	50	2292.0
			10	62.02	100	9579.5
1.75	2.104	3.392	1	2.17	15	149.3
			3	0.01	30	740.9
			5	4.92	50	2229.6
			10	52.11	100	9451.5
2.00	3.019	4.870	1	6.25	15	132.3
			3	0.25	30	702.3
			5	2.25	50	2162.3
			10	42.25	100	9312.3

[†]If $p/h > \rho(.20)$ and $2K/h\mu_b \geq (2K/h\mu_b)_{CR}$, then $s^* < 0$ and $s^a = s^*$, $\Delta^a = \Delta^*$. If $p/h > \rho(.20)$ and $2K/h\mu_b < (2K/h\mu_b)_{CR}$, then $s^* > 0$ and $s^a = s^*$, $\Delta^a = \Delta^*$ provided that $\Delta^a/\mu_b > \rho(.10)$.

Table 5.5

Performance of Asymptotic Approximations for Periodic Review
Systems $(L = 0)$

Dist.	α	n	RE^a (%)			RE^r (%)		
			Max.	Ave.	Min.	Max.	Ave.	Min.
GA	All	216	1.31	0.02	0.00	6.21	0.98	0.00
WL	All	216	3.59	0.05	0.00	20.22	2.28	0.00
TN	All	216	3.66	0.08	0.00	11.02	1.31	0.00
IG	≤ 12.0	180	2.75	0.16	0.00	16.34	1.34	0.00
LN	≤ 1.0	108	1.24	0.04	0.00	14.33	1.55	0.00
All above[†]		936	3.66	0.07	0.00	20.22	1.49	0.00
IG[#]	20.0	36	8.16	2.94	0.00	30.68	7.75	0.00
LN[#×]	1.5, 2.0	72	34.20	2.46	0.00	27.44	5.28	0.19
LN[#+]	3.0	36	434.13	97.55	1.50	102.12	32.99	7.17

[a]Asymptotic approximation from (5.26)-(5.27) and (5.29)-(5.30).

[r]Revised power approximation from (5.36)-(5.37).

[†]Cases for which accuracy conditions (5.32)-(5.33) are satisfied.

[#]Cases for which accuracy conditions are generally not satisfied.

[×]Asymptotic approximation failed ($\Delta^a < 0$) in 3 cases.

[+]Asymptotic approximation failed in 10 cases.

and their relative errors are given in Appendix 2 for some parameter configurations.

Condition (5.39) was violated in 84 (out of 1,080) cases. Most of these (60) occurred under LN, $\alpha = 2, 3$. Other violations were under LN, $\alpha = 1.5$, $p/h = 1, 2$; WL, $\alpha = 0.6$, $p/h = 1$; and IG, $\alpha = 0.5$, $p/h = 1$. Most of these latter are rather extreme cases that are unlikely to arise in practice. When condition (5.39) held, so that Δ_1 was large enough, (5.40) correctly predicted the sign of s^* in all except 40 cases. In all the failures, the inequality in (5.40) held as $\leq$ while $s^* > 0$. Most failures were inconsequential, however, in that both $s^*(> 0)$ and $s^a(< 0)$ were close to zero and the accuracy of (s^a, Δ^a) was not measurably affected. (In 32 out of 40 failures we found $s^* < 0.5$; and, $s^* = 1.43 (> 1)$ in only one case). When condition (5.39) was satisfied and (5.40) held as $>$, indicating large Δ_1 and $s^* > 0$, the accuracy condition (5.41) was satisfied in all but 38 cases. Most of the violations (29) occurred under IG, $\alpha \geq 8$ and LN, $\alpha = 2.0$. Thus both accuracy conditions were satisfied and condition (5.40) was successful in most of the cases we examined. Violations were confined to larger demand coefficients of variation than about 2.0 and to extreme parameter configurations.

In almost all cases satisfying the accuracy conditions, the asymptotic approximation (s^a, Δ^a), computed from (5.26)-(5.27) and (5.29)-(5.30), was practically the same as the optimal policy. In these cases, the revised power approximation (s^r, Δ^r) also performed well in terms of the performance measure RE. However, at times (s^r, Δ^r) was considerably different from the optimal policy (see Appendix 2). When $s^* > 0$ and (s^a, Δ^a) is computed from (5.29)-(5.35), so that (given s^a) Δ^a is distribution-free, the asymptotic approximation was again extremely accurate under the same conditions. However, when p/h was large (i.e., ≥ 10) and K/h was small (i.e., ≤ 10), (5.29)-(5.35) sometimes failed to produce a solution with $\Delta^a > 0$; this was encountered in 28 cases.

In conclusion, the approximations we examined in this section are highly accurate when $c_b < 2.0$. When the accuracy conditions given above hold, (s^a, Δ^a) is nearly the same as (s^*, Δ^*). When $c_b > 2.0$, the range of applicability is reduced (i.e., LN demand with $\alpha > 1.5$). When c_b is very small and μ_b is large (i.e., IG demand with $\alpha \geq 12.0$) this range is again reduced.

5.2   Positive Lead Times

Computational problems are more serious for $L > 0$, due to the complicated form of the lead time demand distribution $H(L, x)$. Additional simplifications are needed to devise algorithms that are usable in practice. The most prevalent of these simplifications has been the approximation of the lead time demand distribution by a normal distribution. Other distributions (gamma, Weibull, etc.) have also been proposed in the literature as approximating distributions. Use of the normal distribution can be justified on the basis of the central limit theorem in situations where L is sufficiently large. However, the determination of what constitutes a large enough L must again be based on empirical considerations.

In the literature, substitution of a more tractable distribution for the lead time demand distribution is often accompanied by other simplifications involving the measure of effectiveness being used. Most of these are ad hoc adjustments achieved by dropping certain terms that are deemed insignificant. From an algorithmic point of view these types of approximation are often unnecessary as they do not really simplify the computational problem. They could also be counterproductive as pointed out in Zipkin (1986).

In this section, we discuss the computation and validation of approximations that are based on: 1) the replacement of the renewal function (of the underlying demand sequence) by its linear asymptote, and 2) the use of the normal distribution for the lead time demand distribution as an asymptotic approximation for large L. The first represents a continuation of the approach of the previous section; it is expected to yield accurate results under similar conditions. Clearly, this simplification is not required for the unit demands model. The second is based on fitting a normal distribution to the asymptotic mean and variance of the lead time demand distribution for large L. This approach is similar to the one used in Tijms and Groenevelt (1984) for the approximation of the reorder point through service level constraints. In what follows we develop algorithms for the computation of these *asymptotic* approximations, based on the first two moments of the underlying distribution(s), and identify their accuracy conditions. We first discuss the unit demands case. This is followed by the general model that includes both the continuous and the periodic review modes.

5.2.1 Unit Demands

If we denote by μ_h and σ_h the mean and the standard deviation of the stationary distribution of demand during a lead time, we find from (3.77-3.78) that:

$$\mu_h = L/\mu_a \tag{5.42}$$

and

$$\sigma_h = \frac{\sigma_a}{\mu_a} \sqrt{\frac{L}{\mu_a}} \tag{5.43}$$

Note that the coefficient of variation is:

$$c_h = \frac{c_a}{\sqrt{L/\mu_a}} \tag{5.44}$$

where c_a is the interarrival time coefficient of variation. Clearly, for $L > \mu_a$, which should normally hold, the lead time demand is relatively less variable than the interarrival time.

We now approximate the lead time distribution by a normal distribution with mean (5.42) and standard deviation (5.43). Appealing to the optimality conditions (4.41-4.42), which are necessary and sufficient, approximations for optimal policies could then be computed through a simple search routine.

Actually, the exact results are not difficult to compute in the unit demands case provided that $H(L, m)$ can be expressed in closed form. For example, if $A(x)$ is Erlang with parameters λ and k (cf. p.d.f. (2.19)), then the lead time demand c.d.f. is seen to be:

$$H(L, m) = \frac{e^{-\lambda L}}{k} \sum_{j=mk}^{(m+1)k-1} \sum_{i=0}^{j} \frac{(\lambda L)^i}{i!}. \tag{5.45}$$

Computationally, the use of this form in (4.41-4.42) is not much more complicated or inefficient than the use of the normal c.d.f. In both cases the search routine could be made very efficient. The difficulty with the exact computations arises when $H(L, m)$ could be computed only by numerical means. For example, if $A(x)$ is one of the distributions considered in the previous section, the convolutions of $A(x)$ and therefore $H(L, m)$ must be computed numerically. There is also the additional consideration that the form of the interarrival distribution may not be known. The approximations suggested above require only its mean and variance.

Exact and approximate policies are compared in Table 5.6 under a set of parameter settings. The exact results represent gamma interarrival times (cf. (5.45)). It is seen that the approximations are satisfactory even for $L = \mu_a$; and they improve with increasing L/μ_a.

5.2.2   The General Case

We now extend the analysis to continuous review systems under batch demand and to periodic review systems. We need the additional simplification of replacing the renewal function $R(x)$ of the underlying demand sequence by its linear asymptote $x/\mu_b + (c_b^2 - 1)/2$. It can be seen that with this change the cost rate function is unimodal and thus the optimality conditions (4.2-4.3) have a unique solution. (Note that $1 + R(x) = C_b + x/\mu_b$ $(C_b = (1 + c_b^2)/2)$ is concave and therefore log-concave so that the main requirement of unimodality that we discussed in Chapter 4 is met.)

Table 5.6

Optimal Policies and Approximations for the Unit Demand Model

L/μ_a	p/h	K/h	$c_a = 5$				$c_a = 1.0$			
			s^*	Δ^*	s^a	Δ^a	s^*	Δ^*	s^a	Δ^a
1	1	1	0	2	1	1	0	2	1	1
		5	-1	4	0	3	-1	4	0	3
		10	-2	6	0	4	-2	6	0	4
		50	-5	13	-4	12	-6	14	-4	12
	2	1	1	1	1	1	1	1	1	2
		5	0	3	1	3	0	3	1	3
		10	0	4	0	5	0	4	0	5
		50	-2	11	-2	11	-3	12	-2	11
	5	1	1	1	2	1	2	1	2	1
		5	1	3	2	2	1	3	2	3
		10	1	4	1	4	1	4	1	5
		50	0	10	0	10	0	10	0	10
	20	1	2	1	2	1	3	1	3	1
		5	2	2	2	2	2	3	3	2
		10	1	4	2	4	2	4	2	4
		50	1	9	2	9	1	10	2	9
2	1	1	1	2	2	1	1	2	2	1
		5	0	4	1	3	0	4	1	4
		10	-1	6	1	4	-1	6	0	5
		50	-5	14	-3	12	-5	14	-3	12
	2	1	2	1	2	1	2	1	3	1
		5	2	2	2	3	1	4	2	3
		10	1	4	1	5	0	6	1	5
		50	-2	12	-1	11	-2	12	-1	11
	5	1	2	1	3	1	3	1	3	2
		5	2	3	3	2	2	3	3	3
		10	2	4	2	4	2	5	3	4
		50	1	10	1	10	1	10	1	11
	20	1	3	1	3	1	4	1	5	1
		5	3	2	3	3	4	3	4	3
		10	3	4	3	4	3	5	4	4
		50	2	10	3	9	3	10	3	10

Table 5.6 (continued)

L/μ_a	p/h	K/h	$c_a=5$				$c_a=1.0$			
			s^*	Δ^*	s^a	Δ^a	s^*	Δ^*	s^a	Δ^a
5	1	1	4	2	5	1	4	2	5	2
		5	3	4	4	4	3	4	4	4
		10	2	6	3	5	2	6	3	6
		50	-1	13	0	12	-2	14	0	12
	2	1	5	1	6	1	5	2	6	2
		5	4	4	5	3	4	4	5	4
		10	4	5	4	5	4	5	4	6
		50	2	11	2	11	2	11	2	12
	5	1	6	1	6	1	7	1	7	2
		5	5	3	6	3	6	4	6	4
		10	5	4	5	5	5	5	6	5
		50	4	10	4	10	4	11	4	11
	20	1	7	1	7	1	9	1	9	1
		5	6	3	7	2	8	3	8	3
		10	6	4	6	5	8	4	8	4
		50	5	10	6	10	6	11	7	10
20	1	1	19	2	20	2	19	2	20	2
		5	18	4	19	4	17	6	18	6
		10	17	6	18	6	16	8	17	8
		50	13	14	15	12	12	16	14	14
	2	1	20	2	21	2	21	2	21	3
		5	20	3	20	4	19	6	21	4
		10	19	5	19	6	19	7	20	6
		50	17	11	17	12	16	14	17	13
	5	1	22	1	22	2	23	2	24	2
		5	21	3	21	4	23	4	23	5
		10	20	5	21	5	22	6	22	6
		50	19	11	19	11	20	12	20	13
	20	1	23	2	24	1	27	2	28	1
		5	23	3	23	3	26	4	26	4
		10	22	5	23	4	26	5	26	5
		50	21	11	22	10	24	12	24	12

For the purposes of this section, we rewrite the optimality conditions as:

$$\frac{H(L,[s]^+)[1 + R(S - [s]^+)] + \int_0^{S-[s]^+} [1 + R(u)]h(L, S - u)du}{1 + R(\Delta)} =$$

$$\frac{p}{p + h} \quad (5.46)$$

$$\frac{p[\Delta + \bar{R}(\Delta)]}{p + h} - \int_0^{S-[s]^+} [1 + R(u)]H(L, S - u)du = \frac{k}{p + h}. \quad (5.47)$$

where $[s]^+ = max(s, 0)$. For the renewal function $R(.)$ and its integral $\bar{R}(.)$ we now use the approximations (5.4) and (5.5). In addition, we regard $H(L, .)$ as the normal c.d.f. and $h(L, .)$ the normal p.d.f., and denote by μ_h and σ_h^2 the corresponding mean and variance. If $\Phi(x)$ and $\phi(x)$ denote the standard normal c.d.f. and p.d.f., then it is well known that:

$$\int_x^\infty u\phi(u)du = \phi(x), \quad (5.48)$$

and

$$\int_x^\infty u^2\phi(u)du = 1 - \Phi(x) + x\phi(x). \quad (5.49)$$

Using these identities, the integral terms in (5.46) and (5.47) can be worked out as:

$$\int_0^{S-[s]^+} [1 + R(u)]h(L, S - u)du \simeq$$

$$(C_b + \frac{S-\mu_h}{\mu_b}[H(L,S) - H(L,[s]^+)] - \frac{\sigma_h}{\mu_b}[h(L,[s]^+) - h(L,S)], \quad (5.50)$$

and

$$\int_0^{S-[s]^+} [1 + R(u)]H(L, S - u)du \simeq$$

$$(C_b + \frac{S-[s]^+}{2\mu_b})(S - [s]^+)H(L,[s]^+) + K_1[H(L,S) - H(L,[s]^+)]+$$

$$K_2[h(L,S) - h(L,[s]^+) - K_3(S - [s]^+)h(L,[s]^+), \quad (5.51)$$

where

$$K_1 = C_b(S - \mu_h) + \frac{(S - \mu_h)^2 + \sigma^2}{2\mu_b}, \tag{5.52}$$

$$K_2 = C_b\sigma_h + \frac{\sigma_h(S - \mu_h)}{2\mu_b}, \tag{5.53}$$

and

$$K_3 = \frac{\sigma_h}{2\mu_b}. \tag{5.54}$$

The mean and the variance of the stationary distribution of the lead time demand are given by (3.47) and (3.48). For L large enough, if we approximate $\bar{R}(L)$ as before, we find the following approximation for the variance:

$$\sigma_h^2 \simeq \frac{L}{\mu_a^2}(L\sigma_b^2 + \frac{\sigma^2 + \mu_b^2}{\mu_a}\sigma_a^2). \tag{5.55}$$

The coefficient of variation is:

$$c_h \simeq c_b\sqrt{1 + \frac{c_a^2(1 + c_b^2)}{L/\mu_a}}. \tag{5.56}$$

An algorithm for the simultaneous iterative solution of (5.46)-(5.47) is tested under a limited range of parameter settings. The exact policies for test cases were computed using exponential, gamma and mixed-exponential distributions for interarrival times and batch sizes (cf. Şahin and Kilari, 1984). A comprehensive assessment of the accuracy of these approximations is not yet available. However, it appears that the accuracy is reasonably good when $c_h < 1$, but may deteriorate very rapidly with increasing c_h for $c_h > 1$. Note that $c_h < 1$ only if $c_b < 1$. Some examples are presented in Table 5.7.

Finally, we note that the approximations described above are also applicable to the periodic review system. As before, the only adjustments needed are $\mu_a = 1$, $\sigma_a^2 = 0$ and $L + 1$ (integer) instead of L. We then have $\sigma_h = (L + 1)\sigma_b$ and $c_h = c_b$.

Table 5.7

Optimal Policies and Approximations Under Batch Demands
$(\mu_a = \mu_b = 5, \, h = 1)$

							$(c_a, \quad c_b, \quad c_h)$				
L	p	K	s^*	Δ^*	s^a	Δ^a		s^*	Δ^*	s^a	Δ^a
				$(1.00, 0.71, 0.87)$					$(0.80, 1.20, 1.48)$		
15	2	10	12.8	8.4	16.1	8.4		13.0	6.4	18.3	8.1
		30	10.8	12.9	13.5	13.6		11.6	9.0	16.5	11.4
		50	9,6	15.7	11.9	16.8		9.2	12.9	13.6	17.1
	5	10	20.5	8.7	24.8	8.6		21.1	6.1	30.3	6.9
		30	18.4	11.8	22.6	11.8		19.9	8.4	28.8	9.6
		50	17.2	14.6	21.4	14.9		17.7	13.1	26.3	14.6
				$(1.00, 0.71, 0.87)$					$(0.80, 1.20, 1.48)$		
25	2	10	23.8	9.2	28.3	10.0		23.9	31.4	32.4	42.4
		30	21.5	14.1	25.4	15.9		21.4	33.5	28.9	45.6
		50	20.1	17.1	23.6	19.6		19.8	35.3	26.9	47.5
	5	10	32.9	8.4	41.1	8.7		34.1	40.8	51.0	59.7
		30	30.8	12.9	38.6	14.2		31.7	43.0	48.0	62.4
		50	29.6	15.9	37.1	16.6		30.4	44.8	46.3	64.4

Chapter 6

Extensions

In this chapter, we consider the operating characteristics of three extensions of the basic model: (1) systems where an order may arrive in two shipments, (2) systems with ordering delays, and (3) systems with random lead times.

6.1 Two-shipment Order Arrivals

In some applications, orders may not be received in full in a single installment after a given lead time. This could be due to the inability or unwillingness of the supplier to fill an order completely with one shipment, or to an attempt to improve service by shipping an order in two installments. A follow-up shipment may also be used to replace the defectives found in the first shipment. Defectives could be produced by the manufacturing processes of the supplier, or these could be items damaged (or lost) in transit.

Moinzadeh and Lee (1989) consider a system like the one described above under the assumption of unit demands generated by a Poisson process. Here we present a generalization of their model to an arbitrary interarrival time distribution, while retaining the unit-demands assumption. Specifically, we assume that a random part, say O, of an order of size $\Delta + 1$ is received after a lead time of L. (Recall that in the unit demands case our convention is to order $\Delta + 1$ when the inventory position reaches down to $s - 1$.) Let $e(j)$, $j = 0, 1, ..., \Delta + 1$, denote the conditional probability mass function of this first installment given that the order size is $\Delta + 1$. The balance, $\Delta + 1 - O$, is then received after an additional delay of ℓ time units. We retain the other assumptions of the basic continuous review model, including full backlogging and the notation introduced before.

It should be noted that the inventory position (= on hand, plus on order minus backorders) is unaffected by the above complication in order deliveries. Order points are still regeneration points of the $\{I_p(t), \quad t \geq 0\}$ process; and the marginal distributions of $I_p(t)$ and I_p remain to be (3.74) and (3.75). However, the relationship between inventory on hand and inventory position is now more complicated. Following Moinzadeh and Lee (1989), this relationship can be expressed as follows. (As before, we denote by $D(t, t + u) \equiv N(t, t + u)$ the quantity demanded during $(t, t + u]$, where t is not necessarily an arrival point; and write $N(u) \equiv N(t, t + u)$ if t is an arrival point.)

$$I(t + L + \ell) = \begin{cases} I_p(t) - D(t, t + L + \ell) & , \text{ if } I_p(t) - D(t, t + \ell) \geq s \\ \\ I_p(t) - D(t, t + L + \ell) \\ \qquad + V(t, t + \ell) & , \text{ if } I_p(t) - D(t, t + \ell) < s \end{cases}$$

$$(6.1)$$

where $V(t, t + u)$ represents the units received through the first shipments of the orders placed during $(t, t + u]$. This relationship is easily verified by noting that if $I_p(t) - D(t, t + \ell) \geq s$, so that no order is placed during $(t, t + \ell]$, then the top relationship holds. For any order placed before t would have been delivered in full by $t + L + \ell$, and no part of any order placed after $t + \ell$ would have been delivered by $t + L + \ell$. In the contrary case, there will be one or more orders placed during $(t, t + \ell]$. And the first shipment of these, but not the second shipments, would have arrived by $t + L + \ell$.

The random variable $V(t, t + \ell)$ can be expressed as

$$V(t, t + \ell) = \sum_{i=1}^{M} O_i \tag{6.2}$$

where O_i, $i = 1, 2, ..., M$ are the i.i.d. first shipments received of orders placed during $(t, t + \ell)$ and $M \equiv M(t, t + \ell)$ is the number of such orders. Since the order size is the same $(= \Delta + 1)$ in each order, this number is:

$$M(t, t + \ell) = \lfloor \frac{S - I_p(t) + D(t, t + \ell)}{\Delta + 1} \rfloor \tag{6.3}$$

where $\lfloor x \rfloor$ represents the integer part of x. Note that the numerator in (6.3) is the total deficit (with respect to S) in inventory position during $(t, t+\ell]$. The random variables that define $I(t)$ by (6.1) are not independent except when the demand process is Poisson; this special case is examined in Moinzadeh and Lee (1989). For the general case, we need the joint distribution:

$$\gamma \equiv \gamma(t, u, v; n, k, i)$$
$$= P[N(t) = n, N(t, t + u) = k, N(t + u, t + u + v) = i],$$
$$t, u, v \geq 0, n, k, i \geq 0. \tag{6.4}$$

This is obtained by an obvious extension of Lemma 2.1 as follows.

$$\gamma = \int_{x=0}^{t} a_n(x) \int_{y=0}^{u} a(t - x + y) \int_{z=0}^{u-y} a_{k-1}(z)$$

$$\int_{w=0}^{v} a(u - y - z + w) \left[A_{i-1}(v - w) - A_i(v - w) \right]$$

$$dw \, dz \, dy \, dx, \quad n \geq 0, \quad k \geq 1, \quad i \geq 1, \tag{6.5a}$$

$$\gamma = \int_{x=0}^{t} a_n(x) \int_{w=0}^{v} a(t - x + u + w)$$

$$\left[A_{i-1}(v - w) - A_i(v - w) \right] dw \, dx,$$

$$n \geq 0, \quad k = 0, \quad i \geq 1, \tag{6.5b}$$

$$\gamma = \int_{x=0}^{t} a_n(x) \int_{y=0}^{u} a(t - x + y) \int_{z=0}^{u-y} a_{k-1}(z)$$

$$[1 - A(u - y - z + v)] \, dz \, dy \, dx,$$
$$n \geq 0, \quad k \geq 1, \quad i = 0, \qquad (6.5c)$$

$$\gamma = \int_{0}^{t} a_n(x) \left[1 - A(t - x + u + v)\right] dx,$$
$$n \geq 0, \quad k = 0, \quad i = 0, \qquad (6.5d)$$

where $a_0(x) = 0$ if $x > 0$ and $a_0(0) = 1$.

The joint distributions of $I_p(t)$, $D(t, t + \ell)$ and $D(t + \ell, t + \ell + L)$ can now be written down by conditioning on the renewal epochs of the process $\{I_p(t), \ t \geq 0\}$. For $t \geq 0$, $\ell \geq 0$, $L \geq 0$, and $n, k, i \geq 0$, we define:

$$q(t, \ell, L; n, k, i) = P[I_p(t) = n, \ D(t, t + \ell) = k, \ D(t + \ell, t + \ell + L) = i]$$
$$(6.6)$$

and let

$$q(n, k, i) \equiv P[I_p = n, \ \tilde{D}(\ell) = k, \ \tilde{D}(\ell, \ell + L) = i]$$

$$= \lim_{t \to \infty} q(t, l, L; n, k, i). \qquad (6.7)$$

For the time-dependent distribution, we have:

$$q(t, \ell, L; n, k, i) = \ \gamma(t, \ell, L; S - n, k, i)$$
$$+ \int_{0}^{t} m(t - u)\gamma(u, \ell, L; S - n, k, i) \, du \qquad (6.8)$$

where $m(u)$ is the renewal density of inventory cycles defined by

(3.72). On passing to the limits, and writing $q \equiv q(n, k, i)$, we obtain:

$$
q = \begin{cases}
\frac{1}{\mu_a[\Delta+1]} \int_{y=0}^{\ell} [1 - A(y)] \\
\quad \int_{z=0}^{\ell-y} a_{k-1}(z) \int_{w=0}^{L} a(\ell - y - z + w) \\
\quad [A_{i-1}(L - w) - A_i(L - w)] \, dw \, dz \, dy, \\
\qquad\qquad k \geq 1, \quad i \geq 1, \\[2ex]
\frac{1}{\mu_a[\Delta+1]} \int_{w=0}^{L} [1 - A(\ell - w)] \\
\quad [A_{i-1}(L - w) - A_i(L - w)] \, dw, \\
\qquad\qquad k = 0, \quad i \geq 1, \\[2ex]
\frac{1}{\mu_a[\Delta+1]} \int_{y=0}^{\ell} [1 - A(y)] \int_{z=0}^{\ell-y} a_{k-1}(z) \\
\quad [1 - A(L + \ell - y - z)] \, dz \, dy, \\
\qquad\qquad k \geq 1, \quad i = 0, \\[2ex]
\frac{1}{\mu_a[\Delta+1]} \int_0^{\infty} [1 - A(u + L + \ell)] \, du, \\
\qquad\qquad k = 0, \quad i = 0.
\end{cases} \qquad (6.9)
$$

Since $\{I_p(t), \ t \geq 0\}$ process is not affected by order splitting, the stationary distribution of inventory position remains to be the uniform distribution $1/(1 + \Delta)$ over $\{s, s + 1, ..., S\}$ (cf. Corollary 3.8). In view of this, (6.9) implies that $I_p(t)$ and $(D(t, t + \ell), D(t + \ell, t + \ell + L))$ are asymptotically independent, as anticipated also by Corollary 3.5.

The relationship (6.1) and the joint distributions (6.8) and (6.9) can now be used to construct the time dependent and stationary distributions of on hand inventory. For the former, we have, for

$j \leq S$:

$$P[I(t + L + \ell) = j] =$$
$$\sum_{n,k,i} P[I(t + L + \ell) = j | I_p(t) = n, D(t, t + \ell) = k, D(t + \ell, t + \ell + L)$$
$$= i] P[I_p(t) = n, D(t, t + \ell) = k, D(t + \ell, t + \ell + L) = i]$$

$$(6.10)$$

In view of (6.1), we can write:

$$P[I(t + L + \ell) = j] = \sum_{n=j \vee s}^{S} \sum_{k=0}^{n-j \vee s} q(t, \ell, L; n, k, n - j - k)$$

$$(6.11)$$

$$+ \sum_{n=s}^{S} \sum_{k=n-s+1}^{\infty} \sum_{i=i_1}^{i_2} e_{m*}(j - n + k + i) q(t, \ell, L; n, k, i)$$

where $m^* = \lfloor (k + S - n)/(\Delta + 1) \rfloor$, $e_{m*}(.)$ is the m^*th convolution of $e(.)$, $i_1 = [n - j - k]^+$, and $i_2 = S - j - mod_{\Delta+1}(k + S - n)$. $(j \vee s = max(j, s)$, $[x]^+ = max(x, 0)$, $\lfloor x \rfloor$ =greatest integer contained in x, $mod_x(y)$=remainder of y divided by x.) The lower limit on i is so that the argument of $e_{m*}(.)$ is nonegative. The upper limit on i expresses the requirement that the sum of m^* first shipments cannot contain more than $m^*(\Delta + 1)$ units. That is: $j - n + k + i \leq \lfloor (k + S - n)/(\Delta + 1) \rfloor (\Delta + 1) = k + S - n - mod_{\Delta+1}(k + S - n)$. On passing to the limits in (6.11) and rewriting the second sum of the second term in a more convenient form, we find the stationary distribution of inventory on hand to be:

$$f(j) = \sum_{n=j \vee s}^{S} \sum_{k=0}^{n-j \vee s} q(n, k, n - j - k)$$

$$(6.12)$$

$$+ \sum_{n=s}^{S} \sum_{m=0}^{\infty} \sum_{k=k_1}^{k_2} \sum_{i=i_1}^{i_2} e_{m+1}(j - n + k + i) q(n, k, i), \quad j \leq S.$$

where $q(n, k, i)$ are given by (6.9), $k_1 = n + m(\Delta+1) - s + 1$, $k_2 = n + (m+1)(\Delta+1) - j \vee s$, $i_1 = [n-j-k]^+$, and $i_2 = (m+1)(\Delta+1) + n - j - k$.

As suggested in Moinzadeh and Lee, one possible application of the model is to situations where the first shipment is inspected for quality, and any defectives found are replaced by the manufacturer through a second shipment. If the proportion of defective items turned out by the manufacturing processes of the supplier is p, then, given $\Delta + 1$, the number of items accepted of the first shipment will have a binomial distribution with parameters $1 - p$ and $\Delta + 1$. Assuming independence, the total number accepted in m first shipments will also have a binomial distribution with parameters $1 - p$ and $m(\Delta + 1)$; that is:

$$e_m(i) = \binom{m(\Delta + 1)}{i}(1 - p)^i p^{m(\Delta + 1) - i}, \quad m \geq 1, \quad 0 \leq i \leq m(\Delta + 1).$$

(6.13)

with this specification, the model can be used to relate quality control to inventory policy.

6.2 Ordering Delays

In this section we consider an extension of the basic model where there is a difference between the time an order is indicated (i.e. , $I_p(t) < s$) and the time the order is actually placed. This difference may be significant in some cases due to delays caused by setup times for a production process or to non-routine order processing that may include vendor selection. More specifically, we will allow for a delay of random length before the placement of each order. Such delays will be assumed to be mutually independent random variables, with common distribution $E(x)$, $x \geq 0$, that are also independent of the demand process.

Without further assumptions, the complication introduced above destroys the simple regenerative structure of the $\{I_p(t), \quad t \geq 0\}$ process. First, since the inventory position may be further depleted during an order delay, it may not be restored to S when the order is finally placed. In order to avoid this difficulty, we will assume that at the time of the placement of an order, its size is adjusted to cover any additional inventory depletion during the delay. Thus at the time it is placed, an order raises the inventory position to S. Second, because of random delays that operate independently of the demand process, points of order placement will not coincide with demand points. As a result, each inventory cycle will start with an incomplete interval (a forward recurrence time) relative to the demand incidence process. In order to get back to a renewal process of cycles, we will also assume that the inter arrival time distribution is negative exponential; thus $A(x) = 1 - e^{-\lambda x}$, $x \geq 0$. Both assumptions are introduced by Weiss (1988) who studied the lost-sales version of the model (i.e. , no backlogging) under the additional assumptions of unit demands and immediate deliveries following order placements. In this section , we allow for backordering, non-zero delivery lead times and arbitrarily distributed demand batch sizes. (For convenience, we shall retain the notation $A(.)$ for the interarrival time distribution, with the understanding that it now represents the exponential distribution.)

To characterize the inventory position, let $I_p(0) = S$ and define:

$$U_1 = inf\{t: \ I_p(t) < s\} \tag{6.14}$$

and

$$T_1 = U_1 + W_1 \tag{6.15}$$

where W_1 is an order delay time with c.d.f. $E(.)$. Evidently, an order is indicated at time U_1 and placed at time T_1 where T_1 is the length of the first cycle. In general, let U_n be the waiting time until the next order-trigger point, measured from an order placement point, and $T_n = U_n + W_n$ the length of the $n-th$ cycle. Under the assumptions made, $\{U_n, \ n = 1, 2, ...\}$ and $\{T_n, \ n = 1, 2, ...\}$ are those of i.i.d. random variables. We have

$$P[U_n \leq t] = 1 - \sum_{k=0}^{\infty} [A_k(t) - A_{k+1}(t)]B_k(\Delta), \tag{6.16}$$

and

$$P[T_n \leq t] = \int_0^t P[U_n \leq u] \ dE(t - u). \tag{6.17}$$

It follows that

$$E[T_n] = \mu_e + \mu_a[1 + R(\Delta)], \tag{6.18}$$

where $\mu_e = E[W_n]$, $\mu_a = \lambda^{-1}$, and, as before, $R(.)$ is the renewal function of the renewal process defined by demand batch sizes. Recall also that V_Δ is the *forward recurrence time at time* Δ of this renewal process. The joint density function, $\ell_\Delta(x, y)$, of U_1 and V_Δ is given by (2.40).

Consider now the renewal process $\{T_n, \ n = 1, 2, ...\}$. Let $\tilde{m}(t), \ t \geq 0$, denote its renewal density, as characterized by the interval distribution (6.17). It can be seen that

$$P[I_p(t) = S] = 1 - A(t) + \int_0^t \tilde{m}(t - u)[1 - A(u)] \, du, \qquad (6.19)$$

and, the density function of $I_p(t)$ is:

$$f_p(t, x) = \begin{cases} \displaystyle\sum_{n=1}^\infty [A_n(t) - A_{n+1}(t)]b_n(S - x) \\ \quad + \displaystyle\sum_{n=1}^\infty b_n(S - x) \int_0^t \tilde{m}(t - u)[A_n(u) - A_{n+1}(u)]du, \\ \hfill s \leq x < S, \\[2mm] \displaystyle\int_{u=0}^t [1 - E(t - u)] \sum_{n=0}^\infty [A_n(t - u) - A_{n+1}(t - u)] \\ \qquad \displaystyle\int_{y=0}^{s-x} b_n(s - y - x)\ell_\Delta(u, y)dy \, du \\ \quad + \displaystyle\int_{y=0}^t \tilde{m}(t - u) \int_{u=0}^y [1 - E(y - u)] \\ \qquad \displaystyle\sum_{n=0}^\infty [A_n(y - u) - A_{n+1}(y - u)] \\ \qquad \displaystyle\int_{z=0}^{s-x} b_n(s - x - z)\ell_\Delta(u, z)dz \, du \, dy, \quad x < s, \end{cases}$$

$$\hfill (6.20)$$

where $b_0(x) = 0$ if $x > 0$ and $b_0(0) = 1$. The expression for $x < s$ follows from the observation that, given $U_n = u$ and $V_\Delta = y$, $0 \leq u \leq t$, $0 \leq y \leq s - x$, for example, we should have a total demand of about $s - y - x$, resulting from $n \geq 0$ withdrawals, and no order placement during $t - u$. The inventory position at time t will then be about $x < s$. Note that for $n = 0$, the first term of the expression for $x < s$, for example, reduces to:

$$\int_0^t \ell_\Delta(u, s - x)[1 + A(t - u)][1 - E(t - u)]du,$$

to account for the event: $\{u < U^{(n)} \leq u+du, \ s-x < V_\Delta \leq s-x+dx,$ no demand during $(0, t - u]$, no order placement during $(0, t - u]\}$.

On passing to the limits, the stationary distribution of the inventory position is obtained as:

$$P[I_p = S] = \frac{\mu_a}{E[T_n]} \tag{6.21}$$

and

$$f_p(x) = \begin{cases} \dfrac{\mu_a r(S - x)}{E[T_n]}, & s \leq x < S \\[3mm] \dfrac{1}{E[T_n]} \displaystyle\sum_{n=0}^{\infty} \int_{y=0}^{\infty} [1 - E(y)][A_n(y) - A_{n+1}(y)]dy \\[3mm] \qquad\qquad \displaystyle\int_{z=0}^{s-x} b_n(s - x - z)v_\Delta(z)dz, & x < s, \end{cases} \tag{6.22}$$

where $v_\Delta(z)$, the forward recurrence time density, is given by (2.16), and $E[T_n]$ by (6.18). Since $A(.)$ is the exponential distribution, $\mu_a = \lambda^{-1}$, and the expression for $x < s$ reduces to

$$f_p(x) = \frac{1}{E[T_n]} \sum_{n=0}^{\infty} \int_{y=0}^{\infty} [1 - E(y)]e^{-\lambda y}\frac{(\lambda y)^n}{n!}b_{n+1}(s - x)dy, \quad x < s. \tag{6.23}$$

As an example, consider the case of unit demands (i.e., a Poison demand process) which implies $r(.) = 1$, and $b_n(j) = 1$ if $n = j$. Further, let $P[W_n = 0] = \pi$ and $P[W_n = w] = 1 - \pi$. Thus, a proportion, $1 - \pi$, of orders are delayed by w units of time. Then, $\mu_e = w(1 - \pi)$, and we find:

$$P[I_p = j] = \begin{cases} \dfrac{1}{1+\Delta+\lambda w(1-\pi)}, & s \leq j < S \\[4mm] \dfrac{1-\pi}{1+\Delta+\lambda w(1-\pi)}[1 - \displaystyle\sum_{i=0}^{s-j-1} e^{-\lambda w}\frac{(\lambda w)^i}{i!}], & j \leq s - 1. \end{cases} \tag{6.24}$$

Note that, as opposed to the case in section 6.1 of multiple shipments, the ordering delays do not complicate the relationship between inventory position and inventory on hand. As for the basic model, the two processes are still related by (3.20), and, under the assumption of a compound Poisson demand process, the results of Section 3.2 could easily be adopted to the present case. In particular, the stationary distribution of on hand inventory is given by the density function (3.43), with $f_p(x)$ as in (6.23) and

$$h(L,y) = \sum_{k=1}^{\infty} e^{-\lambda L}\frac{(\lambda L)^k}{k!}b_k(y). \qquad (6.25)$$

(Note that $P[I = S] = e^{-\lambda L}P[I_p = S]$.) With this modification, the analysis could be extended to measures of effectiveness and optimization problems.

6.3 Random Lead Times

Lead time unreliability is known to have a critical impact on stocking levels and costs (cf. Gross and Soriano 1969). In spite of this, analytical and optimization results are generally lacking for inventory systems with random lead times, due to certain modeling difficulties. As opposed to the constant lead times case, there is no guarantee that all the outstanding orders will arrive in the same sequence as they are placed. This rules out any simple relationship between inventory position and on hand inventory; and the random lead time case cannot be treated as a simple generalization of the constant lead time model. Put differently, if the lead times are regarded as independent random variables (i. e. , the multi-supplier case) then orders can cross; if the orders are not allowed to cross (i.e., the single-supplier case) then the lead times cannot be independent.

Discussion of the continuous review model with random lead times goes back to Karlin and Scarf (1958), Scarf (1958), and Galliher, Morse and Simond(1959). These earlier studies were based largely on analogies to queuing theory and limited to exponential lead times and

Poisson demand processes. In later studies that did not rely heavily
on queuing, the modeling dilemma was circumvented by assumptions.
Hadley and Within (1963), for example, studied the impact of random
lead times on a range of continuous and periodic review models under
the assumption that lead times are independent but that orders do
not cross. This assumption was also used by Kaplan (1970), Urbach
(1977), and others. Another way out of the dilemma was suggested
by Washburn (1973) who assumed that items are not interchangable,
thus decoupling the orders so that it becomes immaterial whether
they cross or not.

In this section, we provide a brief account of the continuous review model under random lead times. We assume that lead times
are i.i.d. random variables that are also independent of the demand
process. The approach we use is a generalization of Finch (1961)
who provided the earliest exact analysis of the unit demands case
without artificial assumptions. We confine our attention to the characterization of the time dependent and limiting behavior of on hand
inventory. We retain the previous notation and conventions with few
exceptions. We assume that the demand batch sizes are discrete with
c.d.f. $B(j)$, $j = 0, 1, ...$, and p.d.f. $b(j)$, $j = 0, 1, ...$. Also, although
the control policy is again an (s, S) policy, in view of discrete demand
batches we now use the convention that an order is placed to raise
the inventory position to S whenever $I_p(t) \leq s$ (rather than $I_p(t) < s$,
as before). The procurement lead time distribution is $L(x)$, $x \geq 0$.
Finally, instead of on hand inventory $I(t)$, we focus on the inventory
deficit $S - I(t)$; we define:

$$p(t, j) = P[S - I(t) = j], \quad t \geq 0, \quad j = 0, 1, ..., \qquad (6.26)$$

and

$$p(j) = \lim_{t \to \infty} p(t, j), \quad j = 0, 1, ... \qquad (6.27)$$

As before, let $D(t)$ be the total demand during $(0, t]$, immediately
following a withdrawal; thus:

$$h(t, k) \equiv P[D(t) = k] = \sum_{n=0}^{\infty} [A_n(t) - A_{n+1}(t)] b_n(k), \quad t \geq 0, \quad k = 0, 1, ...$$

$$(6.28)$$

Time points at which the inventory position is raised to S again form a renewal process with interval (cycle) distribution.

$$C(x) \equiv P[T_n \leq x] = \sum_{n=1}^{\infty} A_n(x)[B_{n-1}(\Delta - 1) - B_n(\Delta - 1)], \quad x \geq 0,$$

(6.29)

where $\Delta = S - s \geq 1$. It follows that $E[T_n] = \mu_a[1 - R(\Delta - 1)]$ where $R(j) = \sum_{n=1}^{\infty} B_n(j)$, $j = 0, 1, ...,$ is the renewal function of the renewal process formed by discrete batch sizes. The joint density of V_Δ, the forward recurrence time at time Δ, of this renewal process and the cycle length T_n is:

$$\ell_\Delta(x, k) = \sum_{j=1}^{\infty} a_j(x) \sum_{i=0}^{\Delta-1} b_{j-1}(i) b(\Delta - i + k), \quad x \geq 0, \quad k = 0, 1, ...,$$

(6.30)

which is a version of (2.35) for discrete batch size distribution. Moments of the forward recurrence time distribution at Δ are given, by Lemma 2.2, as:

$$E[V_\Delta^n] = \sum_{j=0}^{\infty}(j - \Delta)^n b(j) + \sum_{i=1}^{\Delta} r(\Delta - i)[\sum_{j=0}^{\infty}(j - i)b(j) - (-i)^r],$$

$$n = 1, 2, ...$$

(6.31)

We denote the n-th binomial moment of the distribution of V_Δ by $\xi^{(n)}$; we have:

$$\xi^{(n)} = \sum_{k=n}^{\infty} \binom{k}{n} P[V_\Delta = k] = \sum_{k=1}^{n} u_{k,n} E[V_\Delta^n], \quad n = 1, 2, ... , \quad (6.32)$$

where $u_{k,n}$ are the Stirling numbers of the second kind. Similarly, we define

$$\eta^{(n)}(x) = \sum_{k=n}^{\infty} \binom{k}{n} \ell_\Delta(x, k), \quad x \geq 0, \quad n = 1, 2, ... \quad (6.33)$$

Following these preliminaries, we now characterize the distribution $p(t, j)$, $t \geq 0$, $j = 0, 1, ...,$ of inventory deficit at time t. By conditioning on the first renewal epoch of the renewal process formed by inventory cycles, we can see that these probabilities satisfy the system of equations:

$$p(t,j) = h(t,j) + \int_0^t L(t-x)p(t-x,j)dC(x), \quad j = 0,1,...,\Delta-1,$$

(6.34)

$$p(t,j) = \int_0^t L(t-x)p(t-x,j)dC(x)$$
$$+ \int_0^t [1-L(t-x)] \sum_{k=0}^{j-\Delta} p(t-x,j-\Delta-k)\ell_\Delta(x,k)dx,$$
$$j = \Delta, \Delta+1, ...$$

(6.35)

If we let

$$\Pi(t,z) = \sum_{j=0}^\infty z^j p(t,j),$$

and

$$\Lambda(t,z) = \sum_{j=0}^\infty z^j \ell_\Delta(t,j),$$

we obtain from (6.34) and (6.35) that

$$\Pi(t,z) = \sum_{j=0}^{\Delta-1} z^j h(t,j) + \int_0^t L(t-x)\Pi(t-x,z)dC(x)$$
$$+ z^\Delta \int_0^t [1-L(t-x)]\Pi(t-x,z)\Lambda(x,z)dx.$$

(6.36)

For the limiting distribution of inventory deficit, we define

$$\Pi(z) \equiv \sum_{j=0}^\infty z^j p(j) = \lim_{t\to\infty} \Pi(t,z).$$

(6.37)

Explicit results can be obtained for the limiting distribution, if the lead time distribution is a mixture of exponentials (cf. form (2.21)). As an example, for $b(j) = \rho(1-\rho)^{j-1}$, $j = 1,2,...$ (in which case

$r(j) = \rho$ and $R(j) = j\rho$) and $L(x) = 1 - e^{-\lambda x}$, it turns out that

$$\Pi(z) = \frac{1}{1 + (\Delta - 1)\rho}[1 + \frac{\rho z(1 - z^{\Delta-1})}{1 - z}]$$
$$+ \frac{1}{\mu_a[1 + \rho(\Delta - 1)]} \sum_{j=1}^{\infty}(-1)^j[1 - \frac{\rho z^{\Delta}}{1 - z(1 - p)}]^j K_j(z),$$

$$(6.38)$$

where

$$K_j(t) = [1 + \frac{\rho z[1 - (zw(j\lambda))^{\Delta-1}]}{1 - zw(j\lambda)}]A_j, \qquad (6.39)$$

$$A_j = \frac{1 - \alpha(j\lambda)}{j[1 - \alpha(j\lambda)w(j\lambda)^{\Delta-1}]} \prod_{i=1}^{j-1} \frac{\alpha(i\lambda)w(i\lambda)^{\Delta-1}}{1 - \alpha(i\lambda)w(i\lambda)^{\Delta-1}}, \qquad (6.40)$$

$$w(\theta) = 1 - \rho + \rho\alpha(\theta), \qquad (6.41)$$

and $\alpha(\theta) = \int_0^{\infty} e^{-\theta x}a(x)dx$ is the Laplace transform of the interarrival time density $a(x)$. (For details and additional examples, see Şahin, 1983.) On inverting (6.38), we obtain the stationary distribution of inventory deficit as follows.

$$p(0) = \frac{1}{\mu_c}[\mu_a + \sum_{j=1}^{\infty}(-1)^j D_{j,0}], \qquad (6.42a)$$

$$p(k) = \frac{\rho}{\mu_c}[\mu_a + \sum_{j=1}^{\infty}(-1)^j D_{j,k-1}], \quad k = 1, 2, ..., \Delta - 1, \qquad (6.42b)$$

$$p(k) = \frac{1}{\mu_c}[\sum_{i=1}^{n}(-\rho)^i(1 - \rho)^{(n-i)\Delta}$$
$$(\begin{array}{c}(n - i)\Delta + i - 1 \\ (n - i)\Delta\end{array}) \sum_{j=i}^{\infty}(-1)^j(\begin{array}{c}j \\ n\end{array})D_{j,0}$$
$$+ \sum_{i=1}^{n-1}(-\rho)^{i+1} \sum_{r=(n-i-1)\Delta+1}^{(n-1)\Delta-1} (\begin{array}{c}i + r - 1 \\ r\end{array})(1 - \rho)^r$$
$$\sum_{j=1}^{\infty}(-1)^j(\begin{array}{c}j \\ i\end{array})D_{j,(n-i)\Delta-r-1}], \quad k = n\Delta, \ n \geq 1, \qquad (6.42c)$$

$$p(k) = \frac{1}{\mu_c}[\sum_{i=1}^{n}(-\rho)^i(1-\rho)^{k-i\Delta}(\begin{array}{c} k-(\Delta-1)i-1 \\ k-i\Delta \end{array})$$

$$\sum_{j=i}^{\infty}(-1)^j(\begin{array}{c} j \\ n \end{array})D_{j,0}$$

$$+\sum_{i=1}^{n}(-\rho)^{i+1}\sum_{r=(n-i)\Delta}^{k-i\Delta-1}(\begin{array}{c} i+r-1 \\ r \end{array})$$

$$(1-\rho)^r\sum_{j=i}^{\infty}(-1)^j(\begin{array}{c} j \\ i \end{array})D_{j,n-i\Delta-r-1}],$$

$$n\Delta < k < (n+1)\Delta, \quad n \geq 1, \quad (6.42.\text{d})$$

where $\mu_c = \mu_a[1 + \rho(\Delta + 1)]$ and $D_{j,k} = A_j w(j\lambda)^k$.

For the general model with arbitrary batch size and lead time distributions, moments of the time dependent and limiting distributions of inventory deficit can be extracted from the generating function (6.38). For this purpose, let

$$\gamma^{(n)}(t) = \sum_{k=n}^{\infty}(\begin{array}{c} k \\ n \end{array})p(t,k), \quad t \geq 0, \quad n = 1, 2, ..., \qquad (6.43)$$

and

$$\gamma^{(n)}(t) = \sum_{k=n}^{\infty}(\begin{array}{c} k \\ n \end{array})p(k), \quad n = 1, 2, ..., \qquad (6.44)$$

denote the respective binomial moments. On account of:

$$\gamma^{(n)}(t) = \frac{1}{n!}\frac{d^n \Pi(t,z)}{dz^n}|_{z=1}, \qquad (6.45)$$

it follows from (6.36) for $n = 1, 2, ..., \Delta - 1$ that

$$\gamma^{(n)}(t) = \int_0^t \gamma^{(n)}(t-x)dC(x) + \sum_{j=n}^{\Delta-1}(\begin{array}{c} j \\ n \end{array})h(t,j)$$

$$+ \int_0^t[1 - L(t-x)]\sum_{j=0}^{n-1}\gamma^{(j)}(t-x)\sum_{m=0}^{n-j}(\begin{array}{c} \Delta \\ m \end{array})\eta^{(n-j-m)}(x)dx \qquad (6.46)$$

Given $\gamma^{(j)}(t)$, $j = 0, 1, ..., n - 1$ ($\gamma^{(0)}(t) \equiv 1$), this is a renewal equation on $\gamma^{(n)}(t)$. By Lemma 2.5, it has the solution given by:

$$\gamma^{(n)}(t) = \delta^{(n)}(t) + \int_0^t \delta^{(n)}(x)m(t - x)dx \qquad (6.47)$$

where

$$\delta^{(n)}(t) = \sum_{j=n}^{\Delta-1} \binom{j}{n} h(t, j) + \int_0^t [1 - L(t - x)] \sum_{j=0}^{n-1} \gamma^{(j)}(t - x)$$
$$\sum_{m=0}^{n-j} \binom{\Delta}{m} \eta^{(n-j-m)}(x)dx, \quad n = 1, 2, ..., \Delta - 1,$$

$$(6.48)$$

and $m(.)$ is the renewal density defined by inventory cycles with c.d.f. (6.29). A similar development determines $\delta^{(n)}(t)$ for $n \geq \Delta$ as:

$$\delta^{(n)}(t) = \int_0^t [1 - L(t - x)] \sum_{j=0}^{n-1} \gamma^{(j)}(t - x)$$
$$\sum_{m=0}^{\Delta \wedge (n-j)} \binom{\Delta}{m} \eta^{(n-j-m)}(x)dx, \quad n = \Delta, \Delta + 1, ...$$

$$(6.49)$$

Binomial moments of the limiting distribution of inventory deficit follow from these to be:

$$\gamma^{(n)} = \begin{cases} \dfrac{\mu_a}{\mu_c} \displaystyle\sum_{j=n}^{\Delta-1} \binom{j}{n} r(j) + \dfrac{1}{\mu_c} \displaystyle\sum_{j=0}^{n-1} \sum_{m=0}^{n-j} \binom{\Delta}{m} \xi^{(n-j-m)} \\ \displaystyle\int_0^\infty [1 - L(x)]\gamma^{(j)}(x)dx, \ n = 1, 2, ...\Delta - 1, \\[2ex] \dfrac{1}{\mu_c} \displaystyle\sum_{j=0}^{n-1} \sum_{m=0}^{\Delta \wedge (rn-j)} \binom{\Delta}{m} \xi^{(n-j-m)} \\ \displaystyle\int_0^\infty [1 - L(x)]\gamma^{(j)}(x)dx, \ n = \Delta, \Delta + 1, ... \end{cases} \qquad (6.50)$$

where $\mu_c = \mu_a[1 + R(\Delta - 1)]$.

References

1. Archibald, B.C., and Silver, E.A. (1978), "(s,S) Policies Under Continuous Review and Discrete Compound Poisson Demand, " *Management Science*, 24, 889-909.

2. Arrow, K.J., Harris, T., Marschak, J. (1951), "Optimal Inventory Policy, " *Econometrica*, 19, 250-272.

3. Arrow, K.J., Karlin, S., Scarf, H. (eds.) (1958), *Studies in the Mathematical Theory of Inventory and Production*, Stanford University Press.

4. Avriel, M. (1976), *Nonlinear Programming, Analysis and Methods* , Prentice-Hall.

5. Barlow, R.E., Marshall, A.W., and Proschan, F. (1963), "Properties of Probability Distributions With Monotone Hazard Rate, " *Annals of Mathematical Statistics*, 34, 375-389.

6. Barlow, R.E., and Proschan, F. (1975), *Statistical Theory of Reliability and Life Testing: Probability Models*, Rinehart and Winston.

7. Baxter, L.A., Scheuer, E.M., Blischke, W.R., and McConalogue, D.J. (1982a), "On the Tabulation of the Renewal Function, " *Technometrics*, 24, 151-156.

8. Baxter, L.A., Scheuer, E.M., Blischke, W.R., and McConalogue, D.J. (1982b), *Renewal Tables: Tables of Functions Arising in Renewal Theory* , Graduate School of Business Administration, University of Southern California.

9. Beckman, M. (1961), "An Inventory Model for Arbitrary Interval and Quantity Distribution of Demand, " *Management Science*, 8, 35-57.

10. Brown, M. (1980), "Bounds, Inequalities and Monotonicity Properties for Some Specialized Renewal Processes, " *The Annals of Probability*, 8, 227-260.

11. Carlsson, H. (1983), "Remainder Term Estimates of the Renewal Function, " *The Annals of Probability*, 11, 143-157.

12. Çınlar, E. (1975), *Introduction to Stochastic Processes*, Prentice Hall.

13. Coleman, R. (1982), "The Moments of Forward Recurrence Time, " *European Journal of Operations Research*, 9, 181-183.

14. Cléroux, R., and McConalogue, D.J. (1976), "A Numerical Algorithm for Recursively Defined Convolution Integrals Involving Distribution Functions, " *Management Science*, 22, 1138-1146.

15. Cox, D.R. (1962), *Renewal Theory* , Methuen.

16. Cox, D.R. and Smith, W. (1961), *Queues*, Methuen.

17. Ehrhardt, R. (1979), "The Power Approximation for Computing (s,S) Inventory Policies, " *Management Science*, 25, 777-786.

18. Ehrhardt, R., and Kastner, G. (1980), "An Empirical Comparison of Two Approximately Optimal (s,S) Inventory Policies, " Technical Report 16, School of Business Administration, University of North Carolina, Chapel Hill.

19. Ehrhardt, R., and Mosier, C. (1984), "A Revision of the Power Approximation for Computing (s,S) Inventory Policies, " *Management Science*, 30, 618-622.

20. Federgruen, A. and Zipkin, P. (1984), "An Efficient Algorithm for Computing Optimal (s,S) Policies, " *Operations Research*, 32, 1268-1285.

21. Feller, W. (1966), *An Introduction to Probability Theory and its Applications*, Vol. 2, Wiley.

22. Finch, P.D. (1961), "Some Probability Theorems in Inventory Control, " *Publication Mathematich Debrecen*, 8, 241-261.

23. Freeland, J., and Porteus, E. (1980), "Evaluating the Effectiveness of a New Method of Computing Approximately Optimal (s,S) Inventory Policies, " *Operations Research*, 28, 353-364.

24. Galliher, H.P., Morse, P.M., and Simond, M. (1959), "Dynamics of Two Classes of Continuous Review Inventory Systems, " *Operations Research*, 7, 362-384.

25. Gross, D., Harris, C., and Roberts, P. (1972), "Bridging the Gap Between Mathematical Inventory Theory and the Construction of a Workable Model, " *International Journal of Production Research*, 10, 201-214.

26. Gross, D., and Ince, R. (1975), "A Comparison and Evaluation of Approximate Continuous Review Inventory Models, " *International Journal of Production Research*, 13, 9-23.

27. Gross, D., and Soriano, A. (1969), "The Effect of Reducing Lead Time on Inventory Levels–Simulation Analysis, " *Management Science*, 16, B61-B76.

28. Hadley, G., and Whitin, T. (1963), *Analysis of Inventory Systems* , Prentice-Hall.

29. Heyman, D.P., and Sobel, M.J. (1984), *Stochastic Models in Operations Research*, Vol. II, McGraw-Hill.

30. Iglehart, D. (1963), "Optimality of (s,S) Policies in the Infinite Horizon Dynamic Inventory Problem, " *Management Science*, 9, 259-267.

31. Kaplan, R.S. (1970), "A Dynamic Inventory Model With Stochastic Lead Times, " *Management Science*, 16, 491-507.

32. Karlin, S., and Scarf, H. (1958), "Inventory Models and Related Stochastic Processes, " pp. 319-336 in Arrow, Karlin, Scarf (eds.), *Studies in the Mathematical Theory of Inventory and Production*, Stanford Univ. Press.

33. Kruse, W. K. (1981), "Waiting Time in a Continuous Review (s,S) Inventory System with Constant Lead Times, " *Operations Research*, 29, 202-207.

34. Mangasarian, O . (1965), "Pseudo-Convex Functions, " *SIAM J. Control*, Ser. A, 3, 281-290.

35. McConalogue, D.J. (1978), "Convolution Integrals Involving Probability Distribution Functions (Algorithm 102), " *Computer Journal*, 21, 270-272.

36. McConalogue, D.J. (1981), "An Algorithm and Implementing Software for Calculating Convolution Integrals Involving Distributions with a Singularity at the Origin, " Delft University of Technology, Department of Mathematics and informatics, Report 81-03.

37. Naddor, E. (1975), "Optimal and Heuristic Decisions in Single and Multi-Item Inventory Systems, " *Management Science*, 11, 1236-1249.

38. Moinzadeh, K., and Lee, H. (1989), "Approximate Order Quantities and Reorder Points for Inventory Systems Where Orders Arrive in Two Shipments, " *Operations Research*, 37, 277-287.

39. Naddor, E. (1980), "An Analytic Comparison of Two Approximately Optimal (s,S) Inventory Policies, " Technical Report 330, Dept. of Math. Sci., Johns Hopkins University, Baltimore.

40. Nahmias, S. (1976), "On the Equivalence of Three Approxi-

mate Continuous Review Inventory Models, " *Naval Research Logistics Quarterly*, 23, 31-38.

41. Neuts, M.F. (1981), *Matrix-Geometric Solutions in Stochastic Models: An Algorithmic Approach*, The Johns Hopkins University Press.

42. Porteus, E . (1985), "Numerical Comparison of Inventory Policies for Periodic Review Systems, " *Operations Research*, 33, 134-152.

43. Richards, F.R. (1975), "Comments on the Distribution of Inventory Position in a Continuous-Review (s,S) Inventory System," *Operations Research*, 23, 366-371.

44. Roberts, D. (1962), "Approximations to Optimal Policies in a Dynamic Inventory Model, " in *Studies in Applied Probability and Management Science*, K. Arrow, S. Karlin, H. Scarf (eds.), Stanford University Press.

45. Şahin, İ. (1979), "On the Stationary Analysis of (s,S) Inventory Systems with Constant Lead Times, " *Operations Research*, 27, 717-729.

46. Şahin, İ. (1982), "On the Objective Function Behavior in (s,S) Inventory Models, " *Operations Research*, 30, 709-724.

47. Şahin, İ. (1983a), "On the Continuous Review (s,S) Inventory Model Under Compound Renewal Demand and Random Lead Times, " *Journal of Applied Probability*, 20, 213-219.

48. Şahin, İ. (1983b), "On Sufficient Conditions for Cost Rate Function Unimodality in Periodic Review (s,S) Inventory Models, " *Operations Research Letters*, 2, 77-79.

49. Şahin, İ. (1986), "On Approximating the Renewal Function With Its Linear Asymptot: How Large is Large Enough?", *Operations*

152

Research Letters, 4, 207-211.

50. Şahin, İ. (1988a), "Optimality Conditions for Regenerative Inventory Systems Under Batch Demands," *Applied Stochastic Models and Data Analysis*, 4, 173-183.

51. Şahin, İ. (1988b), "On Unit-Demand Inventory Systems," Working Paper.

52. Şahin, İ., and Kilari, P. (1984), "Performance of an Approximation to Continuous Review (s,S) Policies Under Compound Renewal Demand," *International Journal of Production Research*, 22, 1027-1032.

53. Şahin, İ., and Sinha, D. (1987a), "On Asymptotic Approximations for (s,S) Policies," *Stochastic Analysis and Applications*, 5, 189-212.

54. Şahin, İ. and Sinha, D. (1987b), "Renewal Approximation to Optimal Order Quantity for a Class of Continuous Review Inventory Systems," *Naval Research Logistics*, 34, 655-667.

55. Scarf, H.E. (1958), "Stationary Operating Characteristics of an Inventory Model with Time Lag," pp. 298-310 in Arrow, Karlin, Scarf (eds.), *Studies in the Mathematical Theory of Inventory and Production*, Stanford Univ. Press.

56. Scarf, H.E. (1960), "The Optimality of (s,S) Policies in the Dynamic Inventory Problem," in K.J. Arrow, S. Karlin, and P. Suppes (editors), *Mathematical Methods in the Social Sciences*, Stanford Univ. Press.

57. Sinha, D. (1985), *Study of Optimal Policy Approximations for a Class of Single Item Inventory Control Problems*, Ph.D. Degree Thesis, The University of Wisconsin-Milwaukee.

58. Sivazlian, B.D. (1984), "A Continuous Review (s,S) Inven-

tory System With Arbitrary Interarrival Distribution Between Unit Demand, " *Operations Research*, 22, 65-71.

59. Smith, W. (1955), "Regenerative Stochastic Processes, " *Proceedings of the Royal Statistical Soc.*, Series A, 6-31.

60. Stidham, S., Jr. (1974), "Stochastic Clearing Systems, " *Stochastic Processes and Their Applications*, 2, 85-113.

61. Stidham, S., Jr. (1977), "Cost Models for Stochastic Clearing Systems, *Operations Research*, 25, 100-127.

62. Stidham, S., Jr. (1986), "Clearing Systems and (s,S) Inventory Systems With Nonlinear Costs and Positive Lead Times, " *Operations Research*, 34, 276-280.

63. Tijms, H.C. (1972), *Analysis of (s,S) Inventory Models*, Mathematical Centre Trachts 40, Mathematich Centrum, Amsterdam.

64. Tijms, H.C. and Gronevelt, H. (1986), "Approximations for (s,S) Inventory Systems With Stochastic Lead Times and a Service Level Constraint, " *European Journal of Operations Research*, 17, 175-192.

65. Urbach, R. (1977), "Inventory Average Costs: Non-Unit Order Size and Random Lead Times, " DRC Inventory Research Office, Army Logistics Management Center, Ft. Lee, Virginia.

66. Veinott, A.F., Jr. (1966), "On the Optimality of (s,S) Inventory Policies: New Conditions and a New Proof, " *SIAM Journal on Applied Mathematics*, 14, 1067-1083.

67. Veinott, A.F., Jr., and Wagner, H.M. (1965), "Computing Optimal (s,S) Inventory Policies, " *Management Science*, 11, 525-552.

68. Wagner, H.M. (1975), *Principles of Operations Research*,

Prentice-Hall.

69. Washburn, A.R. (1973), "A Bi-modal Inventory Study with Random Lead Times, " NTIS Report No. AD769404, Naval Postgraduate School, Monterey, California.

70. Weiss, H.J. (1988), "Sensitivity of Continuous Review Stochastic (s,S) Inventory Systems to Ordering Delays, " *European J. of Operations Research*, 36, 174-179.

71. Whitt, W. (1982), "Approximating a Point Process, I: Two Basic Methods, " *Operations Research*, 30, 125-147.

72. Zabel, E. (1962), "A Note on the Optimality of (s,S) Policies in Inventory Theory, " *Management Science*, 9, 123-125.

73. Zipkin, P. (1986), "Inventory Service Level Measures: Convexity and Approximation, " *Management Science*, 32, 975-981.

Appendix 1

Optimal Policies and Approximations for Continuous Review Systems $(L = 0, p \to \infty)$

$K/h\mu_a$	Δ^*	Δ^a	RE	$K/h\mu_a$	Δ^*	Δ^a	RE
\multicolumn{8}{c}{Gamma Batches}							

Gamma Batches

$\alpha = 7.0$

$K/h\mu_a$	Δ^*	Δ^a	RE	$K/h\mu_a$	Δ^*	Δ^a	RE
5	0.00	n.a.		10	0.00	0.00	0.00
20	12.40	12.35	0.00	50	22.32	22.15	0.00
100	33.31	33.20	0.00	200	48.84	48.76	0.00

$\alpha = 4.0$

$K/h\mu_a$	Δ^*	Δ^a	RE	$K/h\mu_a$	Δ^*	Δ^a	RE
5	0.00	0.00	0.00	10	6.23	6.09	0.01
20	10.00	9.90	0.00	50	17.40	17.34	0.00
100	25.72	25.67	0.00	200	37.46	37.42	0.00

$\alpha = 3.0$

$K/h\mu_a$	Δ^*	Δ^a	RE	$K/h\mu_a$	Δ^*	Δ^a	RE
5	0.00	0.00	0.00	10	5.37	5.48	0.01
20	8.83	8.77	0.00	50	15.24	15.20	0.00
100	22.44	2.41	0.00	200	32.60	32.58	0.00

$\alpha = 2.0$

$K/h\mu_a$	Δ^*	Δ^a	RE	$K/h\mu_a$	Δ^*	Δ^a	RE
5	2.75	2.71	0.00	10	4.68	4.64	0.00
20	7.34	7.32	0.00	50	12.58	12.56	0.00
100	18.45	18.44	0.00	200	26.75	26.74	0.00

$\alpha = 1.5.0$

$K/h\mu_a$	Δ^*	Δ^a	RE	$K/h\mu_a$	Δ^*	Δ^a	RE
5	2.43	2.42	0.00	10	4.10	4.08	0.00
20	6.41	6.39	0.00	50	10.94	10.93	0.00
100	16.03	16.03	0.00	200	23.22	23.21	0.00

$\alpha = 0.6$

$K/h\mu_a$	Δ^*	Δ^a	RE	$K/h\mu_a$	Δ^*	Δ^a	RE
5	1.52	1.52	0.00	10	2.56	2.57	0.00
20	4.02	4.03	0.00	50	6.90	6.90	0.00
100	10.12	10.13	0.00	200	10.92	10.92	0.00

Weibull Batches

$\alpha = 7.0$

$K/h\mu_a$	Δ^*	Δ^a	RE	$K/h\mu_a$	Δ^*	Δ^a	RE
5	2.16	2.54	4.79	10	4.02	3.82	0.70
20	5.89	5.62	0.26	50	9.04	9.18	0.01
100	13.19	13.19	0.00	200	18.85	18.86	0.00

$\alpha = 4.0$

$K/h\mu_a$	Δ^*	Δ^a	RE	$K/h\mu_a$	Δ^*	Δ^a	RE
5	2.33	2.48	0.17	10	3.77	3.74	0.00
20	5.52	5.51	0.00	50	9.03	9.02	0.00

$K/h\mu_a$	Δ^*	Δ^a	RE	$K/h\mu_a$	Δ^*	Δ^a	RE
				$\alpha = 3.0$			
5	2.45	2.44	0.00	10	3.70	3.69	0.00
20	5.46	5.45	0.00	50	8.94	8.93	0.00
100	2.85	12.85	0.00	200	18.39	18.39	0.00
				$\alpha = 2.0$			
5	2.38	2.36	0.00	10	3.62	3.61	0.00
20	5.37	5.36	0.00	50	8.84	8.83	0.00
100	12.74	12.74	0.00	200	18.26	18.26	0.00
				$\alpha = 1.5$			
5	2.30	2.27	0.00	10	3.55	3.54	0.00
20	5.32	5.31	0.00	50	8.83	8.82	0.00
100	12.77	12.76	0.00	200	18.33	18.33	0.00
				$\alpha = 0.6$			
5	1.30	n.a.		10	2.58	1.46	5.12
20	4.62	4.04	0.41	50	8.95	8.80	0.01
100	14.00	13.99	0.00	200	16.91	16.91	0.00
			Truncated	Normal	Batches		
				$\alpha = 4.0$			
5	0.00	0.00	0.00	10	5.26	6.56	3.74
20	9.63	10.34	0.40	50	17.84	17.76	0.00
100	26.15	26.08	0.00	200	37.85	37.82	0.00
				$\alpha = 2.0$			
5	3.24	3.12	0.06	10	5.15	5.05	0.01
20	7.28	7.74	0.00	50	13.06	13.04	0.00
100	19.02	18.99	0.00	200	27.42	27.40	0.00
				$\alpha = 1.0$			
5	2.65	2.59	0.02	10	4.14	4.11	0.00
20	6.26	6.23	0.00	50	10.44	10.42	0.00
100	15.14	15.13	0.00	200	21.79	21.79	0.00
				$\alpha = 0.0$			
5	2.16	2.13	0.01	10	3.33	3.32	0.00
20	5.00	4.99	0.00	50	8.29	8.28	0.00
100	11.99	11.99	0.00	200	17.23	17.23	0.00
				$\alpha = -1.0$			
5	1.80	1.79	0.00	10	2.76	2.76	0.00
20	4.11	4.11	0.00	50	6.78	6.78	0.00
100	9.79	9.79	0.00	200	10.04	10.03	0.00
				$\alpha = -2.0$			
5	1.57	1.56	0.00	10	2.37	2.37	0.00
20	3.51	3.51	0.00	50	5.76	5.76	0.00
100	8.29	8.29	0.00	200	11.87	11.87	0.00

$K/h\mu_a$	Δ^*	Δ^a	RE	$K/h\mu_a$	Δ^*	Δ^a	RE
			Inverse	Gaussian	Batches		
				$\alpha = 20.0$			
5	0.00	n.a.		10	0.00	n.a.	
20	0.00	n.a.	50	50	24.65	32.97	5.79
100	45.44	51.87	1.09	200	81.23	78.32	0.09
				$\alpha = 12.0$			
5	0.00	n.a.		10	0.00	n.a.	
20	0.00	0.00	0.00	50	26.88	27.53	0.03
100	42.01	42.06	0.00	200	62.66	62.48	0.00
				$\alpha = 8.0$			
5	0.00	n.a.		10	0.00	0.00	0.00
20	13.28	12.81	0.01	50	23.65	23.42	0.00
100	35.38	35.25	0.00	200	51.99	51.89	0.00
				$\alpha = 4.0$			
5	0.00	0.00	0.00	10	6.21	6.09	0.01
20	9.99	9.90	0.00	50	17.40	17.34	0.00
100	25.71	25.67	0.00	200	37.45	37.42	0.00
				$\alpha = 1.0$			
5	2.06	2.00	0.02	10	3.37	3.36	0.00
20	5.23	5.25	0.00	50	8.93	8.95	0.00
100	13.10	13.11	0.00	200	18.97	18.98	0.00
				$\alpha = 0.5$			
5	1.46	1.36	0.14	10	2.35	2.32	0.00
20	3.65	3.66	0.00	50	6.26	6.28	0.00
100	9.20	9.22	0.00	200	13.36	13.37	0.00
				Lognormal	Batches		
				$\alpha = 0.1$			
5	2.59	2.61	0.00	10	3.99	3.97	0.00
20	5.89	5.88	0.00	50	9.66	9.66	0.00
100	13.91	13.91	0.00	200	19.92	19.92	0.00
				$\alpha = 0.5$			
5	2.43	2.36	0.02	10	3.92	3.90	0.00
20	6.03	6.03	0.00	50	10.22	10.22	0.00
100	14.93		0.00	200	21.59	21.58	0.00
				$\alpha = 1.0$			
5	2.09	n.a.		10	3.64	3.05	0.78
20	5.91	5.56	0.11	50	10.53	10.40	0.01
100	15.81	15.78	0.00	200	23.33	23.34	0.00

$K/h\mu_a$	Δ^*	Δ^a	RE	$K/h\mu_a$	Δ^*	Δ^a	RE
				$\alpha = 1.5$			
5	1.81	n.a.		10	3.33	n.a.	
20	5.64	n.a.		50	10.54	9.01	0.76
100	16.27	15.28	0.14	200	24.55	23.97	0.02
				$\alpha = 4.0$			
5	1.60	n.a.		10	3.05	n.a.	
20	5.35	n.a.		50	10.38	n.a.	
100	16.42	n.a.		200	25.34	21.37	1.01

Appendix 2

Optimal Policies and Approximations for Periodic Review Systems ($L = 0$)

K/h	p/h	S^*	s^*	S^a	s^a	RE^a	S^b	s^b	RE^b	S^r	s^r	RE^r
					Gamma Dem. ($\alpha = 7.0$)							
5	1	9.08	0.83	8.83	1.17	0.12	8.82	1.18	0.14	7.75	-0.05	1.79
	2	10.19	3.31	10.07	3.53	0.07	9.89	3.63	0.18	10.78	2.98	0.25
	5	10.70	5.74	11.73	5.89	0.24	9.63	6.79	1.60	13.60	5.80	1.21
	20†	12.33	8.83	14.19	8.81	1.31	n.a.	n.a.		16.45	8.66	2.57
10	1	11.53	-1.55	11.31	-1.36	0.04				9.06	-2.00	2.31
	2	13.15	1.95	12.83	2.09	0.05	12.79	2.11	0.06	12.64	1.57	0.19
	5	15.14	4.88	14.69	4.96	0.08	14.24	5.08	0.35	15.92	4.85	0.20
	20	17.87	8.07	17.29	8.12	0.11	n.a.	n.a.		19.10	8.03	0.46
50	1	21.79	-11.69	21.68	-11.65	0.00				16.95	-8.04	3.43
	2	24.80	-3.90	24.68	-3.82	0.00				22.24	-2.76	0.73
	5	27.72	2.06	27.60	2.08	0.00	27.58	2.09	0.00	27.00	2.00	0.04
	20	30.97	6.37	30.72	6.38	0.00	29.28	6.52	0.18	31.34	6.34	0.02
200	1	40.46	-30.46	40.40	-30.41	0.00				34.83	-15.58	6.21
	2	46.30	-14.65	46.24	-14.60	0.00				42.29	-8.12	2.15
	5	51.59	-2.72	51.53	-2.70	0.00				48.96	-1.45	0.28
	20	55.92	4.52	55.80	4.53	0.00	55.65	4.54	0.00	54.86	4.46	0.02
					Gamma Dem. ($\alpha = 4.0$)							
5	1	6.20	-0.70	6.07	-0.57	0.04				4.63	-1.35	3.66
	2	7.17	1.18	7.02	1.26	0.04	7.00	1.27	0.05	6.93	0.96	0.15
	5	8.50	2.98	8.3	3.03	0.03	7.88	3.20	0.50	9.08	3.10	0.33
	20	10.49	5.29	10.32	5.33	0.03	n.a.	n.a.		11.25	5.27	0.37
10	1	7.99	-2.49	7.89	-2.39	0.01				5.65	-2.83	3.91
	2	9.16	0.16	9.06	0.22	0.01	9.06	0.22	0.01	8.37	-0.12	1.56
	5	10.62	2.34	10.49	2.37	0.01	10.37	2.41	0.04	10.86	2.38	0.03
	20	12.69	4.79	12.57	4.81	0.01	n.a.	n.a.		13.28	4.79	0.14
50	1	15.72	-10.22	15.67	-10.17	0.00				11.72	-7.43	3.85
	2	18.00	-4.25	17.95	-4.22	0.00				15.75	-3.41	0.91
	5	20.20	0.26	20.16	0.27	0.00	20.16	0.27	0.00	19.37	0.21	0.09
	20	22.67	3.49	22.57	3.49	0.00	21.98	3.55	0.05	22.67	3.51	0.01
200	1	29.82	-24.32	29.80	-24.20	0.00				25.46	-13.17	6.16
	2	34.24	-12.37	34.22	-12.36	0.00				31.14	-7.49	2.13
	5	38.25	-3.35	38.22	-3.34	0.00				36.22	-2.41	0.28
	20	41.49	2.09	41.46	2.10	0.00	41.41	2.10	0.00	40.71	2.08	0.01

†Accuracy condition (5.35) does not hold

K/h	p/h	S^*	s^*	S^a	s^a	RE^a	S^b	s^b	RE^b	S^r	s^r	RE^r
					Gamma Dem. ($\alpha = 2.0$)							
5	1	3.83	-1.34	3.80	-1.30	0.01				2.50	-1.83	5.17
	2	4.52	-0.01	4.47	0.00	0.01				4.15	-0.18	0.41
	5	5.48	1.20	5.44	1.22	0.00	5.35	1.24	0.04	5.69	1.35	0.19
	20	7.04	2.90	7.00	2.92	0.00	n.a.	n.a.		7.23	2.90	0.04
10	1	5.10	-2.60	5.07	-2.57	0.00				3.26	-2.89	4.91
	2	5.93	-0.72	5.90	-0.70	0.00				5.21	-0.95	0.76
	5	6.97	0.77	6.93	0.78	0.00	6.91	0.78	0.00	6.99	0.84	0.02
	20	8.54	2.54	8.53	2.55	0.00	n.a.	n.a.		8.71	2.56	0.02
50	1	10.56	-8.06	10.54	-8.04	0.00				7.71	-6.18	3.75
	2	12.17	-3.83	12.15	-3.83	0.00				10.59	-3.30	0.85
	5	13.74	-0.65	13.72	-0.64	0.00				13.18	-0.71	0.08
	20	15.50	1.58	15.49		0.00	15.31	1.61	0.01	15.54	1.65	0.00
200	1	20.53	-18.03	20.52	-18.02	0.00				17.73	-10.29	5.73
	2	23.65	-9.58	23.65	-9.57	0.00				21.79	-6.22	1.90
	5	26.50	-3.20	26.48	-3.20	0.00				25.42	-2.59	0.19
	20	28.79	0.61	28.78	0.62	0.00	28.77	0.62	0.00	28.64	0.62	0.02
					Gamma Dem. ($\alpha = 0.6$)							
5	1	1.61	-1.20	1.61	-1.21	0.00				0.97	-1.60	4.90
	2	1.98	-0.48	1.98	-0.49	0.00				1.89	-0.66	0.49
	5	2.52	0.07	2.53	0.06	0.00	2.53	0.06	0.00	2.76	0.20	0.71
	20	3.59	1.05	3.61	1.04	0.00	n.a.	n.a.		3.62	1.07	0.01
10	1	2.30	-1.90	2.31	-1.91	0.00				1.43	-2.20	3.90
	2	2.75	-0.88	2.76	-0.88	0.00				2.53	-1.10	0.57
	5	3.32	-0.10	3.33	-0.11	0.00				3.54	-0.09	0.16
	20	4.38	0.82	4.38	0.82	0.00	4.17	0.87	0.14	4.51	0.88	0.08
50	1	5.30	-4.90	5.30	-4.90	0.00				4.12	-4.07	2.09
	2	6.18	-2.59	6.19	-2.59	0.00				5.76	-2.44	0.20
	5	7.04	-0.85	7.04	-0.85	0.00				7.23	-0.96	0.08
	20	8.05	0.27	8.07	0.27	0.00	8.06	0.27	0.00	8.56	0.37	0.21
200	1	10.76	-10.36	10.77	-10.37	0.00				10.12	-6.40	4.24
	2	12.48	-5.74	12.48	-5.74	0.00				12.43	-4.09	1.23
	3	14.03	-2.25	14.03	-2.25	0.00				14.49	-2.03	0.09

K/h	p/h	S^*	s^*	S^a	s^a	RE^a	S^b	s^b	RE^b	S^r	s^r	RE^r
					Weibull Dem. ($\alpha = 7.0$)							
5	1	2.74	-1.22	2.60	-1.21	0.14				2.47	-0.38	5.63
	2	2.94	-0.31	2.93	-0.30	0.00				2.90	0.05	1.98
	5	3.64	0.37	3.23	0.38	0.53				3.29	0.44	0.76
	20	3.95	0.76	3.47	0.81	1.12	3.40	0.81	1.24	3.63	0.78	0.96
10	1	3.59	-2.12	3.50	-2.11	0.02				3.38	-0.67	7.39
	2	3.91	-0.83	3.97	-0.82	0.01				3.90	-0.15	2.56
	5	4.51	0.14	4.39	0.15	0.05	4.39	0.15	0.06	4.35	0.31	0.45
	20	4.90	0.73	4.70	0.73	0.50	4.68	0.74	0.63	4.76	0.71	0.34
50	1	7.30	-5.90	7.29	-5.90	0.00				7.59	-1.55	14.50
	2	8.36	-3.01	8.34	-3.01	0.00				8.36	-0.77	6.60
	5	9.30	-0.83	9.28	-0.83	0.00				9.05	-0.09	1.66
	20	9.84	0.46	9.91	0.46	0.00	9.91	0.46	0.00	9.64	0.51	0.05
200	1	14.13	-12.74	14.13	-12.74	0.00				15.77	-2.65	20.22
	2	16.26	-6.96	16.25	-6.96	0.00				16.87	-1.56	9.85
	5	18.12	-2.60	18.11	-2.60	0.00				17.84	-0.58	3.04
	20	19.40	-0.01	19.34	-0.01	0.00				18.68	0.26	0.26
					Weibull Dem. ($\alpha = 4.0$)							
5	1	2.55	-1.23	2.53	-1.21	0.00				2.13	-0.69	3.25
	2	2.91	-0.33	2.87	-0.32	0.01				2.68	-0.14	0.78
	5	3.16	0.35	3.19	0.35	0.01	3.18	0.35	0.00	3.17	0.35	0.00
	20	3.38	0.82	3.49	0.82	0.06	3.32	0.84	0.04	3.62	0.80	0.29
10	1	3.43	-2.11	3.42	-2.10	0.00				2.96	-1.05	4.95
	2	3.90	-0.84	3.89	-0.83	0.00				3.61	-0.40	1.53
	5	4.33	0.12	4.32	0.13	0.00	4.32	0.13	0.00	4.19	0.19	0.17
	20	4.69	0.73	4.68	0.74	0.00	4.62	0.74	0.02	4.72	0.71	0.01
50	1	7.15	-5.83	7.15	-5.82	0.00				6.88	-2.16	10.51
	2	8.20	-2.98	8.19	-2.98	0.00				7.86	-1.19	4.42
	5	9.12	-0.83	9.12	-0.83	0.00				8.73	-0.32	0.90
	20	9.78	0.44	9.77	0.44	0.00	9.76	0.44	0.00	9.49	0.44	0.06
200	1	13.88	-12.56	13.88	-12.56	0.00				14.68	-3.56	16.17
	2	15.96	-6.87	15.96	-6.87	0.00				16.06	-2.18	7.56
	5	17.80	-2.57	17.80	-2.57	0.00				17.29	-0.95	2.07
	20	19.04	-0.03	19.03	-0.02	0.00				18.36	0.12	1.21

K/h	p/h	S^*	s^*	S^a	s^a	RE^a	S^b	s^b	RE^b	S^r	s^r	RE^r
					Weibull Dem.	$(\alpha = 2.0)$						
5	1	2.46	-1.25	2.44	-1.23	0.00				1.72	-1.13	3.85
	2	2.81	-0.36	2.80	-0.35	0.00				2.47	-0.38	0.71
	5	3.20	0.32	3.18	0.33	0.00	3.18	0.33	0.00	3.15	0.31	0.02
	20	3.69	0.95	3.66	0.96	0.00	3.17	1.03	1.62	3.79	0.95	0.05
10	1	3.32	-2.11	3.31	-2.10	0.00				2.42	-1.61	3.80
	2	3.80	-0.85	3.79	-0.85	0.00				3.31	-0.72	0.86
	5	4.28	0.09	4.27	0.10	0.00	4.27	0.10	0.01	4.12	0.08	0.03
	20	4.82	0.82	4.80	0.82	0.00	4.63	0.84	0.10	4.85	0.81	0.02
50	1	6.99	-5.78	6.98	-5.77	0.00				5.99	-3.13	6.21
	2	8.03	-2.97	8.02	-2.96	0.00				7.31	-1.80	2.16
	5	9.00	-0.85	8.96	-0.84	0.00				8.50	-0.62	0.28
	20	9.77	0.45	9.72	0.45	0.00	9.71	0.45	0.00	9.55	0.43	0.01
200	1	13.64	-12.43	13.64	-12.43	0.00				13.36	-5.03	10.87
	2	15.70	-6.81	15.70	-6.8-	0.00				15.24	-3.15	4.62
	5	17.54	-2.56	17.53	-2.56	0.00				16.91	-1.48	0.95
	20	18.80	-0.04	18.80	-0.04	0.00				18.37	-0.02	0.03
					Weibull Dem.	$(\alpha = 0.6)$						
5	1¶	2.08	-1.49	1.56	-1.62	1.03				1.56	-2.67	4.38
	2	2.87	-0.48	2.64	-0.60	0.20				3.86	-0.37	2.04
	5	4.46	0.56	4.72	0.38	0.17	4.71	0.38	0.17	6.08	1.85	5.98
	20	8.41	3.81	9.54	3.35	0.76	n.a.	n.a.		8.58	4.35	0.25
10	1¶	2.99	-2.51	2.59	-2.66	0.32				1.87	-4.14	5.96
	2	3.90	-1.02	3.69	-1.13	0.09				4.56	-1.45	0.96
	5	5.49	0.16	5.65	0.07	5.65	0.07	0.02	7.09	1.09	3.79	
	20	9.43	3.21	10.30	2.94	0.34	n.a.	n.a.		3.76	0.34	0.29
50	1¶	7.32	-7.16	7.23	-7.30	0.01				4.92	-8.64	2.90
	2	8.85	-3.61	8.82	-3.69	0.00				8.85	-4.70	0.72
	5	10.70	-0.92	10.78	-0.97	0.00				12.44	-1.12	0.82
	20	14.41	1.61	14.77	1.55	0.02	14.61	1.59	0.00	15.86	2.30	0.71
200	1¶	15.80	-15.81	15.84	-15.91	0.00				13.13	-14.22	0.93
	2	18.58	-8.55	18.65	-8.61	0.00				18.65	-8.69	0.00
	5	21.30	-3.07	21.42	-3.09	0.00				23.62	-3.72	0.53
	20	24.84	0.24	24.96	0.23	0.00	24.96	0.23	0.00	28.13	0.78	0.89

¶Accuracy condition (5.33) does not hold

K/h	p/h	S^*	s^*	S^a	s^a	RE^a	S^b	s^b	RE^b	S^r	s^r	RE^r
Trunc. Normal Dem. $(\alpha = 4.0)$												
5	1	6.21	-0.50	6.20	-0.33	0.03				5.59	-0.28	0.34
	2	7.59	1.40	6.94	1.47	0.98	6.90	1.49	1.09	7.17	1.31	0.42
	5	8.58	2.89	7.73	2.91	2.60	7.50	2.97	4.15	8.61	2.75	0.06
	20	9.57	4.18	8.64	4.19	3.66	n.a.	n.a.		9.97	4.11	0.67
10	1	8.25	-2.32	8.10	-2.22	0.03				7.02	-1.31	2.52
	2	8.78	0.32	9.11	0.38	0.27	9.11	0.38	0.27	8.90	0.57	1.12
	5	9.07	2.36	10.11	2.38	0.84	10.03	2.40	0.77	10.59	2.27	1.17
	20	9.93	3.91	11.15	3.91	1.93	9.28	4.11	1.79	12.15	3.82	2.51
50	1	16.06	-10.15	15.94	-10.10	0.00			14.29	-4.51	5.88	
	2	18.21	-4.18	18.17	- 4.15	0.00				17.09	-1.70	1.96
	5	20.23	0.33	20.17	0.34	0.00	20.17	0.34	0.00	19.60	0.80	0.20
	20	21.81	3.11	21.75	3.11	0.00	21.57	3.13	0.01	21.82	3.02	0.02
200	1	30.17	-24.29	30.14	-24.26	0.00				29.40	-8.51	11.02
	2	34.54	-12.34	34.52	-12.32	0.00				33.36	-4.55	4.69
	5	38.44	-3.31	38.41	-3.31	0.00				36.90	-1.01	0.98
	20	41.25	2.05	41.11	2.05	0.00	41.09	2.05	0.00	39.98	2.07	0.04
Trunc. Normal Dem. $(\alpha = 2.0)$												
5	1	4.08	-1.23	4.04	-1.17	0.01				2.96	-1.32	3.36
	2	4.69	0.12	4.61	0.16	0.02	4.61	0.17	0.02	4.29	0.00	0.44
	5	5.38	1.27	5.28	1.29	0.03	5.20	1.31	0.09	5.50	1.21	0.05
	20	6.21	2.39	6.10	2.40	n.a.	n.a.			6.66	2.38	0.43
10	1	5.41	-2.53	5.36	-2.49	0.00				3.91	-2.18	3.52
	2	6.16	-0.62	6.12	-0.60	0.00				5.47	0.61	0.62
	5	6.98	0.87	6.92	0.88	0.00	6.90	6.88	0.01	6.89	0.81	0.03
	20	7.92	2.16	7.83	2.12	0.00	6.82	2.30	1.51	8.21	2.13	0.09
50	1	10.97	-8.11	10.95	-8.09	0.00				8.90	-4.84	4.93
	2	12.57	-3.82	12.55	-3.81	0.00				11.22	-2.51	1.50
	5	14.04	-0.59	14.02	-0.59	0.00				13.31	-0.42	0.17
	20	15.29	1.49	15.30	1.50	0.00	15.21	1.50	0.00	15.17	1.44	0.00
200	1	21.10	-18.22	21.09	-18.22	0.00				19.53	-8.16	8.85
	2	24.25	-9.66	24.24	-9.66	0.00				22.83	-4.87	3.54
	5	27.06	-3.19	27.05	-3.19	0.00				25.76	-1.93	0.63
	20	29.06	0.66	29.08	0.66	0.00	29.08	0.66	0.00	28.34	0.64	0.04

K/h	p/h	S^*	s^*	S^a	s^a	RE^a	S^b	s^b	RE^b	S^r	s^r	RE^r
				Trunc. Normal Dem. ($\alpha = 0.0$)								
5	1	2.23	-1.26	2.21	-1.24	0.00				1.40	-1.36	4.94
	2	2.59	-0.41	2.58	-0.40	0.00				2.25	-0.51	0.81
	5	3.05	0.25	3.02	0.26	0.00	3.02	0.26	0.00	3.03	0.27	0.00
	20	3.67	1.04	3.65	1.04	0.00	2.81	1.23	5.99	3.78	1.01	0.07
10	1	3.04	-2.07	3.03	-2.06	0.00				2.02	-1.91	4.08
	2	3.52	-0.87	3.50	-0.87	0.00	3.02	-0.91	0.81			
	5	4.03	0.03	4.02	0.03	0.00	4.02	0.03	0.00	3.93	0.01	0.03
	20	4.69	0.87	4.69	0.87	0.00	4.45	0.91	0.19	4.78	0.86	0.01
50	1	6.51	-5.54	6.50	-5.53	0.00				5.25	-3.62	4.36
	2	7.50	-2.87	7.50	-2.86	0.00				6.74	-2.13	1.20
	5	8.42	-0.85	8.42	-0.85	0.00				8.08	-0.79	0.09
	20	9.27	0.41	9.26	0.41	0.00	9.25	0.42	0.00	9.28	0.41	0.00
200	1	12.81	-11.84	12.81	-11.84	0.00				12.13	-5.76	8.00
	2	14.78	-6.50	14.77	-6.50	0.00				14.24	-3.64	3.08
	5	16.53	-2.47	16.53	-2.47	0.00				16.13	-1.76	0.45
	20	17.80	-0.08	17.80	-0.08	0.00				17.79	-0.10	0.00
				Trunc. Normal Dem. ($\alpha = -2.0$)								
5	1	1.42	-1.02	1.42	-1.01	0.00				0.91	-1.02	4.17
	2	1.66	-0.44	1.66	-0.44	0.00				1.44	-0.49	0.74
	5	1.93	-0.01	1.93	-0.01	0.00				1.92	0.00	0.00
	20	2.34	0.44	2.33	0.44	0.00	2.20	0.46	0.23	2.38	0.44	0.01
10	1	1.98	-1.58	1.98	-1.57	0.00				1.38	-1.37	3.39
	2	2.30	-0.76	2.30	-0.76	0.00				2.01	-0.74	0.63
	5	2.62	-0.14	2.62	-0.14	0.00				2.57	-0.17	0.04
	20	4.07	0.23	4.07	0.23	0.00	4.06	0.23	0.00	4.14	0.24	0.03
50	1	4.36	-3.95	4.36	-3.95	0.00				3.76	-2.44	4.70
	2	5.04	-2.13	5.04	-2.13	0.00				4.70	-1.50	1.38
	5	5.65	-0.75	5.65	-0.75	0.00				5.54	-0.66	0.07
	20	6.17	0.07	6.17	0.07	0.00	6.7	0.07	0.00	6.28	0.08	0.05
200	1	8.68	-8.27	8.68	-8.27	0.00				8.73	-3.77	9.11
	2	10.02	-4.62	10.02	-4.62	0.00				10.05	-2.45	3.67
	5	11.21	-1.86	11.21	-1.86	0.00				11.23	-1.27	0.61
	20	12.02	-0.09	12.02	-0.09	0.00				12.16	0.00	0.02

K/h	p/h	S*	s*	S^a	s^a	RE^a	S^b	s^b	RE^b	S^r	s^r	RE^r
				Inverse Gaussian Dem. ($\alpha = 12.0$)								
5	1†	11.77	4.36	12.63	4.87	0.85	12.40	5.10	0.63	12.87	2.76	2.43
	2†	13.05	7.65	14.20	7.76	1.81	13.42	8.21	0.23	16.82	6.71	10.97
	5†	15.18	10.74	16.36	10.71	1.94	n.a.	n.a.		20.50	10.39	16.02
	20†	18.51	14.63	19.67	14.53	1.70	n.a.	n.a.		24.23	14.12	15.57
10	1	16.59	0.89	16.03	1.47	0.11	16.02	1.48	0.12	14.57	0.21	0.72
	2	20.46	5.40	17.98	5.76	1.22	17.75	5.88	1.47	19.23	4.87	0.42
	5†	23.30	9.04	20.40	9.41	1.75	19.11	9.78	4.04	23.50	9.14	0.05
	20†	27.57	13.37	23.87	13.53	2.75	n.a.	n.a.		27.65	13.29	0.00
50	1	30.09	-12.55	29.81	-12.31	0.01				24.75	-7.66	2.67
	2	34.10	-2.24	33.71	-2.11	0.11				31.64	-0.77	0.47
	5	37.44	5.54	37.51	5.60	0.00	37.41	5.62	0.00	37.85	5.43	0.01
	20	41.35	11.15	41.63	11.18	0.00	38.45	11.50	0.51	43.51	11.09	0.18
200	1	54.51	-37.02	54.40	-36.90	0.00				47.89	-17.48	6.04
	2	62.17	-16.34	62.03	-16.27	0.00				57.61	-7.76	2.06
	5	69.10	-0.72	68.94	-0.69	0.00				66.31	0.93	0.23
	20	74.74	8.74	74.53	8.75	0.00	74.20	8.77	0.00	74.01	8.63	0.01
				Inverse Gaussian Dem. ($\alpha = 8.0$)								
5	1	9.93	1.43	9.65	1.84	0.16	9.62	1.88	0.20	8.78	0.46	1.25
	2	11.03	4.03	10.97	4.31	0.11	10.73	4.44	0.25	12.01	3.70	0.24
	5†	11.83	6.58	12.79	6.74	1.29	n.a.	n.a.		15.02	6.71	1.99
	20†	13.50	9.86	15.66	9.95	2.21	n.a.	n.a.		18.08	9.76	2.96
10	1	12.66	-1.11	12.32	-0.82	0.06				10.18	-1.62	1.95
	2	14.33	2.62	13.95	2.78	0.07	13.89	2.81	0.09	14.00	2.19	0.14
	5	16.46	5.68	15.96	5.76	0.10	15.43	5.90	0.48	17.50	5.69	0.36
	20	19.51	9.11	18.93	9.18	0.13	n.a.	n.a.		20.90	9.09	0.55
50	1	23.57	-12.07	23.45	-11.95	0.00				18.58	-8.07	3.28
	2	26.79	-3.65	26.65	-3.57	0.00				24.22	-2.43	0.66
	5	29.94	2.72	29.76	2.75	0.00	29.73	2.75	0.00	29.31	2.66	0.02
	20	33.26	7.26	33.18	7.27	0.00	31.49	7.44	0.22	33.94	7.29	0.01
200	1	43.53	-32.04	43.47	-31.97	0.00				37.63	-16.12	6.18
	2	49.80	-15.14	49.72	-15.11	0.00				45.59	-8.15	2.14
	5	55.46	-2.39	55.36	-2.37	0.00				52.71	-1.03	0.28
	20	60.10	5.30	59.94	5.31	0.00	59.79	5.32	0.00	59.02	5.28	0.02

†Accuracy condition (5.35) does not hold

K/h	p/h	S^*	s^*	S^a	s^a	RE^a	S^b	s^b	RE^b	S^r	s^r	RE^r
						Inverse Gaussian Dem. $(\alpha = 4.0)$						
5	1	6.19	6.07	-0.57	0.03				4.63	-1.35	3.60	
	2	7.16	1.18	7.02	1.24	0.03	7.00	1.25	0.04	6.93	0.96	0.15
	5	8.51	2.87	8.37	2.92	0.02	8.00	3.05	0.33	9.08	3.10	0.41
	20	10.74	5.24	10.63	5.28	0.02	n.a.	n.a.		11.25	5.27	0.15
10	1	7.98	-2.48	7.89	-2.39	0.01				5.65	-2.83	3.90
	2	9.15	0.17	9.06	0.22	0.01	9.06	0.22	0.01	8.37	-0.12	0.39
	5	10.61	2.27	10.51	2.30	0.01	10.41	2.32	0.03	10.86	2.38	0.05
	20	12.93	4.73	12.80	4.75	0.00	n.a.	n.a.		13.28	4.79	0.06
50	1	15.71	-10.21	15.67	-10.17	0.00				11.72	-7.43	3.84
	2	17.99	-4.25	17.95	-4.22	0.00				15.75	-3.41	0.90
	5	20.20	0.26	20.16	0.27	0.00	20.16	0.27	0.00	19.37	0.21	0.09
	20	22.70	3.40	22.66	3.40	0.00	22.11	3.46	0.04	22.67	3.51	0.02
200	1	29.83	-24.32	29.80	-24.30	0.00				25.46	-13.17	6.15
	2	34.24	-12.37	34.22	-12.36	0.00				31.44	-7.49	2.13
	5	38.25	-3.35	38.22	-3.34	0.00				36.22	-2.41	0.28
	20	41.65	2.05	41.47	2.05	0.00	41.44	2.05	0.00	40.71	2.08	0.01
						Inverse Gaussian Dem. $(\alpha = 1.0)$						
5	1	2.35	-1.32	2.35	-1.35	0.00				1.41	-1.77	5.78
	2	2.81	-0.40	2.83	-0.41	0.00				2.60	-0.58	0.50
	5	3.49	0.32	3.53	0.31	0.01	3.53	0.31	0.01	3.70	0.52	0.59
	20	4.89	1.49	4.98	1.48	n.a.	n.a.		4.80	1.62	0.13	
10	1	3.24	-2.22	3.24	-2.24	0.00				1.98	-2.54	4.80
	2	3.82	-0.90	3.83	-0.91	0.00				3.38	-1.14	0.72
	5	4.54	0.09	4.56	0.09	0.00	4.56	0.09	0.00	4.66	0.15	0.06
	20	5.92	1.22	5.95	1.21	0.00	5.38	1.36	0.58	5.90	1.38	0.16
50	1	7.10	-6.09	7.11	-6.11	0.00				5.29	-4.91	2.94
	2	8.24	-3.12	8.25	-3.12	0.00				7.37	-2.83	0.50
	5	9.34	-0.87	9.25	-0.87	0.00				9.23	-0.97	0.03
	20	10.67	0.57	10.67	0.56	0.00	10.65	0.57	0.00	10.93	0.73	0.13
200	1	14.16	-13.15	14.16	-13.16	0.00				12.70	-7.86	4.93
	2	16.36	-7.18	16.37	-7.19	0.00				15.63	-4.94	1.50
	5	18.36	-2.67	18.37	-2.67	0.00				18.25	-2.32	0.08
	20	20.00	0.00	20.00	0.00	0.00	n.a.	n.a.		20.56	-0.01	0.05

K/h	p/h	S^*	s^*	S^a	s^a	RE^a	S^b	s^b	RE^b	S^r	s^r	RE^r
					Lognormal Dem.	($\alpha = 0.1$)						
5	1	2.77	-1.25	2.75	-1.23	0.00				2.22	-0.83	2.89
	2	3.14	-0.28	3.12	-0.27	0.00				2.88	-0.16	0.51
	5	3.50	0.44	3.48	0.45	0.00	3.47	0.45	0.00	3.48	0.43	0.01
	20	3.90	0.98	3.88	0.98	0.00	3.60	1.01	0.42	4.02	0.98	0.05
10	1	3.72	-2.20	3.70	-2.18	0.00				3.06	-1.26	4.11
	2	4.23	-0.83	4.22	-0.82	0.00				3.85	-0.47	1.09
	5	4.70	0.20	4.69	0.21	0.00	4.69	0.21	0.00	4.56	0.23	0.07
	20	5.17	0.87	5.14	0.87	0.00	5.05	0.88	0.33	5.20	0.87	0.00
50	1	7.72	-6.20	7.72	-6.20	0.00				7.16	-2.60	8.85
	2	8.85	-3.14	8.84	-3.13	0.00				8.33	-1.43	3.54
	5	9.85	-0.83	9.85	-0.82	0.00	9.38	-0.38		0.63		
	20	10.5	0.55	10.58	0.55	0.00	10.57	0.55	0.00	10.30	0.54	0.05
200	1	14.97	-13.45	14.97	-13.45	0.00				15.40	-4.128	14.33
	2	17.21	-7.32	17.21	-7.32	0.00				17.06	-2.62	6.53
	5	19.20	-2.70	19.20	-2.69	0.00				18.54	-1.14	1.66
	20	20.53	0.05	20.53	0.05	0.00	20.53	0.05	0.00	19.83	0.15	0.09
					Lognormal Dem.	($\alpha = 0.5$)						
5	1	2.87	-1.34	2.86	-1.35	0.00				1.79	-1.73	5.65
	2	3.38	-0.29	3.38	-0.29	0.00				3.06	-0.46	0.59
	5	4.08	0.55	4.11	0.54	0.00	4.10	0.55	0.00	4.23	0.71	0.31
	20	5.43	1.73	5.53	1.71	0.03	n.a.	n.a.		5.39	1.87	0.15
10	1	3.89	-2.36	3.88	-2.37	0.00				2.45	-2.55	4.82
	2	4.53	-0.86	4.53	-0.87	0.00				3.94	-1.05	0.80
	5	5.29	0.27	5.31	0.27	0.00	5.30	0.27	0.00	5.31	0.31	0.01
	20	6.64	1.46	6.68	1.45	0.00	5.93	1.61	0.82	6.62	1.62	0.16
50	1	8.27	-6.75	8.27	-6.76	0.00				6.20	-5.08	3.56
	2	9.55	-3.38	9.55	-3.38	0.00				8.42	-2.87	0.78
	5	10.78	-0.83	10.78	-0.83	0.00				10.41	-0.87	0.05
	20	12.10	0.80	12.14	0.80	0.00	12.10	0.80	0.00	12.21	0.93	0.06
200	1	16.26	-14.75	16.27	-14.76	0.00				14.51	-8.24	5.95
	2	18.76	-7.99	18.77	-7.99	0.00				17.64	-5.11	2.01
	5	21.02	-2.87	21.02	-2.87	0.00				20.44	-2.32	0.19
	20	22.80	0.16	22.80	0.16	0.00	22.80	0.16	0.00	22.91	0.15	0.00

K/h	p/h	S^*	s^*	S^a	s^a	RE^a	S^b	s^b	RE^b	S^r	s^r	RE^r
					Lognormal Dem. $(\alpha = 1.0)$							
5	1	2.90	-1.42	2.63	-1.58	0.27				1.78	-2.43	6.49
	2	3.61	-0.29	3.53	-0.41	0.08				3.84	-0.38	0.61
	5	4.83	0.73	5.14	0.58	0.24	5.12	0.59	0.22	5.79	1.57	3.74
	20	7.74	2.98	9.01	2.58	1.24	n.a.	n.a.		7.91	3.69	0.79
10	1	3.95	-2.53	3.73	-2.67	0.11				2.24	-3.75	7.05
	2	4.79	-0.90	4.70	-1.00	0.04				4.65	-1.34	0.52
	5	6.05	0.37	6.24	0.29	0.06	6.24	0.29	0.06	6.90	0.90	1.64
	20	8.91	2.51	9.90	2.28	0.53	8.36	2.86	0.44	9.19	3.20	0.76
50	1	8.70	-7.44	8.61	-7.55	0.01				5.74	-7.79	3.22
	2	10.21	-3.67	10.17	-3.73	0.00				9.27	-4.26	0.48
	5	11.87	-0.82	11.92	-0.85	0.00				12.48	-1.05	0.17
	20	14.64	1.32	14.98	1.28	0.03	14.84	1.30	0.01	15.49	1.96	0.66
200	1	17.63	-16.49	17.63	-16.57	0.00				14.46	-12.82	2.20
	2	20.50	-8.86	20.52	-8.91	0.00				19.44	-7.84	0.26
	5	23.20	-3.10	23.26	-3.12	0.00				23.91	-3.38	0.06
	20	26.13	0.33	26.18	0.32	0.00	26.18	0.32	0.00	27.92	0.64	0.32
					Lognormal Dem. $(\alpha = 1.5)$							
5	1¶	2.89	-1.49	1.24	-1.75	8.46				2.16	-2.90	4.85
	2¶	3.79	-2.29	2.89	-0.64	1.97				5.29	0.23	3.94
	5†	5.56	0.94	6.14	0.41	0.86	6.14	0.41	0.85	8.36	3.30	12.04
	20	10.42	4.62	14.47	2.99	4.38	13.35	3.42	2.51	11.98	6.92	2.87
10	1¶	3.96	-2.67	2.43	-2.93	3.57				2.30	-4.88	8.51
	2¶	5.00	-0.92	4.04	-1.22	1.21				5.94	-1.24	0.95
	5†	6.82	0.50	7.10	0.17	0.35	7.10	0.17	0.35	9.42	2.36	7.47
	20	11.73	3.93	15.11	2.78	2.73	14.27	3.06	1.64	13.23	6.04	2.47
50	1¶	8.91	-8.07	7.87	-8.38	0.31				5.29	-10.93	5.62
	2¶	10.70	-3.93	9.92	-4.15	0.18				10.58	-5.64	1.24
	5	12.90	-0.80	12.73	-0.95	0.03				15.43	-0.79	1.14
	20	17.72	2.12	19.40	1.79	0.38	19.19	1.83	0.29	20.15	3.93	2.10
200	1	18.60	-18.25	18.06	-18.57	0.02				14.31	-18.41	1.28
	2	21.89	-9.73	21.47	-9.93	0.02				21.71	-11.01	0.19
	5	25.19	-3.31	25.03	-3.41	0.00				28.39	-4.32	0.78
	20	29.85	0.61	30.49	0.54	0.02	30.49	0.54	0.02	34.52	1.81	1.52

†Accuracy condition (5.35) does not hold

¶Accuracy condition (5.33) does not hold

Index